THE MAHALANOBIS–
TAGUCHI STRATEGY

This book integrates the principles of P. C. Mahalanobis and Genichi Taguchi. This photograph was taken at a rare occasion when these two were together at the Indian Statistical Institute, Calcutta, India, in 1954. Standing from left to right are P. C. Mahalanobis, Pandit Jawaharlal Nehru (then Indian Prime Minister), Walter A. Shewhart (known as the father of quality control), and Genichi Taguchi (shaking hands with the Prime Minister).

THE MAHALANOBIS–TAGUCHI STRATEGY

A Pattern Technology System

GENICHI TAGUCHI
RAJESH JUGULUM

JOHN WILEY & SONS, INC.

This book is printed on acid-free paper. ∞

Copyright © 2002 by John Wiley & Sons, New York. All rights reserved.

Published simultaneously in Canada.

No part of this publication may be reproduced, stored in a retrieval system or transmitted in any form or by any means, electronic, mechanical, photocopying, recording, scanning or otherwise, except as permitted under Sections 107 or 108 of the 1976 United States Copyright Act, without either the prior written permission of the Publisher, or authorization through payment of the appropriate per-copy fee to the Copyright Clearance Center, 222 Rosewood Drive, Danvers, MA 01923, (978) 750-8400, fax (978) 750-4744. Requests to the Publisher for permission should be addressed to the Permissions Department, John Wiley & Sons, Inc., 605 Third Avenue, New York, NY 10158-0012, (212) 850-6011, fax (212) 850-6008, E-Mail: PERMREQ@WILEY.COM.

This publication is designed to provide accurate and authoritative information in regard to the subject matter covered. It is sold with the understanding that the publisher is to engaged in rendering professional services. If professional advice or other expert assistance is required, the services of a competent professional person should be sought.

Wiley also publishes its books in a variety of electronic formats. Some content that appears in print may not be available in electronic books. For more information about Wiley products, visit our web site at www.wiley.com.

Library of Congress Cataloging-in-Publication Data

Taguchi, Genichi, 1924–
    The Mahalanobis-Taguchi strategy : a pattern technology system / Genichi Taguchi, Jugulum Rajesh.
        p. cm.
    ISBN 0-471-02333-7 (acid-free paper)
    1. Taguchi methods (Quality control)   2. Pattern recognition systems.
I. Rajesh, Jugulum. II. Title.
    TS156 .T316 2002
    658.5'62—dc21

                                                                        2002003927

Printed in the United States of America

10 9 8 7 6 5 4 3 2 1

To My Parents, Wife, Brother, and Sisters—Rajesh Jugulum

CONTENTS

Preface   xiii

Acknowledgments   xvii

Terms and Symbols   xix

Definitions of Mathematical and Statistical Terms   xxi

1 Introduction   1

    1.1   The Goal   3

    1.2   The Nature of a Multidimensional System   5

          1.2.1   Description of Multidimensional Systems   5

          1.2.2   Correlations between the Variables   6

          1.2.3   Mahalanobis Distance   6

          1.2.4   Robust Engineering/Taguchi Methods   8

    1.3   Multivariate Diagnosis—The State of the Art   10

          1.3.1   Principal Component Analysis   10

          1.3.2   Discrimination and Classification Method   11

          1.3.3   Stepwise Regression   11

          1.3.4   Test of Additional Information (Rao's Test)   12

          1.3.5   Multiple Regression   12

          1.3.6   Multivariate Process Control Charts   12

          1.3.7   Artificial Neural Networks   13

    1.4   Approach   14

          1.4.1   Classification versus Measurement   14

          1.4.2   Normals versus Abnormals   14

          1.4.3   Probabilistic versus Data Analytic   15

          1.4.4   Dimensionality Reduction   15

1.5   Refining the Solution Strategy   16

1.6   Guide to This Book   16

2   MTS and MTGS   **19**

2.1   A Discussion of Mahalanobis Distance   21

2.2   Objectives of MTS and MTGS   23

    2.2.1   Mahalanobis Distance (Inverse Matrix Method)   24

    2.2.2   Gram–Schmidt Orthogonalization Process   25

    2.2.3   Proof That Equations 2.2 and 2.3 Are the Same   26

    2.2.4   Calculation of the Mean of the Mahalanobis Space   27

2.3   Steps in MTS   30

2.4   Steps in MTGS   31

2.5   Discussion of Medical Diagnosis Data: Use of MTGS and MTS Methods   33

2.6   Conclusions   39

3   Advantages and Limitations of MTS and MTGS   **41**

3.1   Direction of Abnormalities   43

    3.1.1   The Gram–Schmidt Process   44

    3.1.2   Identification of the Direction of Abnormals   44

    3.1.3   Decision Rule for Higher Dimensions   49

3.2   Example of a Graduate Admission System   50

3.3   Multicollinearity   52

3.4   A Discussion of Partial Correlations   55

3.5   Conclusions   57

4   Role of Orthogonal Arrays and Signal-to-Noise Ratios in Multivariate Diagnosis   **59**

4.1   Role of Orthogonal Arrays   62

4.2   Role of *S/N* Ratios   63

4.3   Advantages of *S/N* ratios   71

    4.3.1   *S/N* Ratio as a Simple Measure to Identify Useful Variables   71

    4.3.2   *S/N* Ratio as a Measure of Functionality of the System   79

    4.3.3   *S/N* Ratio to Predict the Given Conditions   80

4.4   Conclusions   81

5   Treatment of Categorical Data in MTS/MTGS Methods   83

    5.1   MTS/MTGS with Categorical Data   85

    5.2   A Sales and Marketing Application   87

        5.2.1   Selection of Suitable Variables   88

        5.2.2   Description of the Variables   88

        5.2.3   Construction of Mahalanobis Space   89

        5.2.4   Validation of the Measurement Scale   89

        5.2.5   Identification of Useful Variables (Developing Stage)   92

        5.2.6   *S/N* Ratio of the System (Before and After)   94

    5.3   Conclusions   95

6   MTS/MTGS under a Noise Environment   97

    6.1   MTS/MTGS with Noise Factors   99

        6.1.1   Treat Each Level of the Noise Factor Separately   101

        6.1.2   Include the Noise Factor as One of the Variables   101

        6.1.3   Combine Variables of Different Levels of the Noise Factor   102

        6.1.4   Do Not Consider the Noise Factor If It Cannot Be Measured   103

    6.2   Conclusions   103

7   Determination of Thresholds—A Loss Function Approach   105

    7.1   Why Threshold Is Required in MTS/MTGS   108

7.2 Quadratic Loss Function 109

    7.2.1 QLF for the Nominal-the-Best Characteristic 109

    7.2.2 QLF for the Larger-the-Better Characteristic 110

    7.2.3 QLF for the Smaller-the-Better Characteristic 111

7.3 QLF for MTS/MTGS 111

    7.3.1 Determination of Threshold 112

    7.3.2 When Only Good Abnormals Are Present 114

7.4 Examples 114

    7.4.1 Medical Diagnosis Case 114

    7.4.2 A Student Admission System 115

7.5 Conclusions 116

8 Standard Error of the Measurement Scale　　　　**117**

8.1 Why Mahalanobis Distance Is Used for Constructing the Measurement Scale 119

8.2 Standard Error of the Measurement Scale 120

8.3 Standard Error for the Medical Diagnosis Example 121

8.4 Conclusions 122

9 Advance Topics in Multivariate Diagnosis　　　　**123**

9.1 Multivariate Diagnosis Using the Adjoint Matrix Method 126

    9.1.1 Related Topics of Matrix Theory 126

    9.1.2 Adjoint Matrix Method for Handling Multicollinearity 128

9.2 Examples for the Adjoint Matrix Method 130

    9.2.1 Example 1 130

    9.2.2 Example 2 136

9.3 β-Adjustment Method for Small Correlations 139

9.4 Subset Selection Using the Multiple Mahalanobis Distance Method 142

    9.4.1 Steps in the MMD Method 145

    9.4.2 Example 145

9.5   Selection of Mahalanobis Space from Historical Data   147

9.6   Conclusions   149

10   MTS/MTGS versus Other Methods   **151**

10.1   Principal Component Analysis   155

10.2   Discrimination and Classification Method   157
    10.2.1   Fisher's Discriminant Function   158
    10.2.2   Use of Mahalanobis Distance   159

10.3   Stepwise Regression   161

10.4   Test of Additional Information (Rao's Test)   163

10.5   Multiple Regression Analysis   165

10.6   Multivariate Process Control   169

10.7   Artificial Neural Networks   169
    10.7.1   Feed-Forward (Backpropagation) Method   170
    10.7.2   Theoretical Comparison   171
    10.7.3   Medical Diagnosis Data Analysis   171

10.8   Conclusions   174

11   Case Studies   **175**

11.1   American Case Studies   177
    11.1.1   Auto Marketing Case Study   177
    11.1.2   Gear-Motor Assembly Case Study   181
    11.1.3   ASQ Research Fellowship Grant Case Study   188
    11.1.4   Improving the Transmission Inspection System
            Using MTS   189

11.2   Japanese Case Studies   191
    11.2.1   Improvement of the Utility Rate of Nitrogen
            While Brewing Soy Sauce   191
    11.2.2   Application of MTS for Measuring Oil in Water
            Emulsion   193
    11.2.3   Prediction of Fasting Plasma Glucose (FPG) from
            Repetitive Annual Health Checkup Data   195

11.3   Conclusions   197

12 Concluding Remarks **199**

    12.1 Important Points of the Proposed Methods 201

    12.2 Scientific Contributions from MTS/MTGS Methods 203

    12.3 Limitations of the Proposed Methods 205

    12.4 Recommendations for Future Research 205

Bibliography **207**

Appendixes **213**

    A.1 ASI Data Set **215**

    **A.2 Principal Component Analysis
          (MINITAB Output)** **217**

    **A.3 Discriminant and Classification Analysis
          (MINITAB Output)** **219**

    **A.4 Results of Stepwise Regression
          (MINITAB Output)** **220**

    **A.5 Multiple Regression Analysis
          (MINITAB Output)** **225**

    **A.6 Neural Network Analysis
          (MATLAB Output)** **226**

    A.7 Variables for Auto Marketing Case Study **227**

Index **229**

PREFACE

Design of a good information system based on several characteristics is an important requirement for successfully carrying out
any decision-making activity. In many cases, although a significant amount of information is available, we fail to use such information in a meaningful way. As we require high-quality
products in day-to-day life, it is also necessary to have high-
quality information systems to make robust decisions or predictions. To produce high-quality products, the variability in the
processes must first be reduced. Variability can be accurately measured and reduced only if we have a suitable measurement system
with appropriate measures. Similarly, in the design of information
systems, it is essential to develop a measurement scale and use
appropriate measures to make accurate predictions or decisions.

Usually, the required information is to be extracted from many
variables (characteristics) that define a multidimensional system.
A multidimensional system could be an inspection system, a medical diagnosis system, a sensor system, a face/voice recognition
system (any pattern recognition system), a credit card/loan approval system, a weather forecasting system, or a university admission system. As we encounter these multidimensional systems
in day-to-day life, it is important to have a measurement scale by
which the degree of abnormality (severity) can be measured. In
the case of medical diagnosis, the degree of abnormality refers to
the severity of a disease; in an inspection system, the degree of
abnormality refers to level of acceptance of a product; and in a
credit card/loan approval system, it refers to the ability to pay
back the balance/loan. A measurement scale based on the char-

acteristics of multidimensional systems greatly enhances the decision maker's ability to make judicious decisions. While developing a multidimensional measurement scale, it is essential to keep in mind the following: (1) having a base or reference point to the scale, (2) validation of the scale, (3) selection of important variables that are adequate to measure the level of abnormality, and (4) future diagnosis with the measurement scale, developed from the important variables.

Several multivariate methods are being used in multidimensional applications, but still there are incidences of false alarms in applications like weather forecasting, air bag sensor operation, and medical diagnosis. These problems could be caused by lack of an adequate measurement system with suitable measures to accurately determine or predict the degree of severity.

This book presents methods to develop a multidimensional measurement scale by integrating mathematical and statistical concepts, such as Mahalanobis distance and the Gram–Schmidt orthogonalization method, with the principles of robust engineering (or Taguchi Methods). The methods are developed by visualizing the multidimensional system in a different way. The measures and procedures used in these methods are data analytic and do not depend on the distribution of the characteristics defining the system. The current trends in multivariate diagnosis/pattern recognition lean toward the data analytic procedures. These methods have been advocated by the authors during recent years. One such method is known as the Mahalanobis–Taguchi system (MTS) and is being used in multidimensional applications.

There is a book on MTS, but it is not complete as it is only a compilation of MTS case studies and does not provide a theoretical basis for MTS. Recently the method of MTS has been refined significantly. This book presents the refined MTS method along with other methods like the Mahalanobis–Taguchi–Gram–Schmidt (MTGS) method and the adjoint matrix method. The book also describes recent real-world problems that are solved using these methods and compares these methods with existing multivariate/pattern recognition techniques.

The book is intended to help the decision makers using multidimensional systems in areas such as pattern recognition, medical diagnosis, inspection, and banking. The methods presented in the book will help them to construct a multivariate measurement scale with which the degree of abnormality can be accurately measured. The methods provide not only a measurement scale but also a means to improve the accuracy of predictions. The book can also be used to teach graduate students who are specializing in the fields of multivariate analysis, robust engineering, design of information systems, and artificial intelligence. Of course, the aim of any learning process is not knowledge alone, but action. The case studies provided in the book enable readers to understand the practical use of these methods, which in turn will enhance their ability to use the methods in their respective fields.

The methods in this book represent one of the perspectives of designing information systems. We hope these methods illustrate how to construct a measurement scale in multidimensional systems, even when there are common problems like multicollinearity and weak correlations, and thereby contribute to the science of measurement in multidimensional applications. In the coming years, the methods provided in the book may become elementary with the introduction of newer methods. If these methods can serve as the foundation for the development of the subject of multivariate measurement science, they will have served a fruitful purpose.

GENICHI TAGUCHI

RAJESH JUGULUM

ACKNOWLEDGMENTS

Of all living things, human beings are the most fortunate because of their ability to think, communicate, plan, and execute actions and offer helping hands to others. It is because of this helping nature that humans derive mutual benefits. Writing this book was possible due to many such helping hands.

First, we thank Shin Taguchi for his continuous encouragement and support right from the beginning. We are thankful to Tatsuji Kanetaka for providing liver disease test data, which was used to illustrate many methods in the book. Since Jugulum's Ph.D. thesis (Jugulum completed his Ph.D. under the guidance of Dr. Taguchi) is related to the proposed methods of this book, thanks are due to the thesis committee members at Wayne State University, Kai Yang, Donald Falkenburg, and Kenneth Chelst, for their ideas and suggestions. We are also grateful to Don Clausing for reviewing the book proposal. His comments helped to improve the contents of this book. We also thank C. R. Rao of Pennsylvania State University and J. K. Ghosh of Indian Statistical Institute for helpful suggestions.

Many people have generously helped by allowing or arranging to get permissions to publish the case studies in the book. Our thanks to all members of the American Supplier Institute (ASI), Louis Lavallee and Steve O'Leyar of Xerox Corporation, members of the Family Fleet Portfolio team (one of the EMMP-2000 leadership teams), John Dickerson and Mahfooz-ul-haq Mian of Ford Motor Company, Mike Massanari of Wayne State University's medical school, Rhonda Lang of the American Society for Quality, Yasuo Ishihara of Ichibiki Company, Kazuyo Tsushita of Aichi Health Plaza, and Yoshio Ishii of Fuji Photo Film Company.

We are very thankful to Robert Argentieri, Bob Hilbert, and Jessica Gallus of John Wiley & Sons for their continued support from the beginning. Finally, we thank S. B. Koimattur for carefully reading the manuscript and providing valuable suggestions.

TERMS AND SYMBOLS

A	Loss corresponding to the distance Λ
A_0	Loss associated with functional limit
α	Statistical level of significance
ANN	Artificial neural networks
β	The estimate of the slope
C	Correlation matrix
C^{-1}	Inverse of the correlation matrix
C_{adj}	Adjoint of the correlation matrix
C_{pooled}	Pooled covariance matrix
D^2	Mahalanobis distance
dB	Decibel units
EMMP	Engineering Management Masters Program
$E(X)$	Expected value of X
F_{in}	Critical F-ratio for an entering variable in stepwise regression
F_{out}	Critical F-ratio for a leaving variable in stepwise regression
GSP	Gram–Schmidt orthogonalization process
K	Number of variables
$L_a\,(b^c)$	Representation of an orthogonal array, where L denotes Latin square design, a = the number of experimental runs, b = the number of levels of each factor, and c = the number of columns in the array
λ	Eigenvalues of principal components
Λ_0	Functional limit
M	Input signal
MATLAB	A mathematical software

MD	Mahalanobis distance
MDA	Distances obtained from an adjoint matrix
m_i	Mean of ith characteristic
MINITAB	A statistical software
MMD	Multiple Mahalanobis distance
MR	Multiple regression
MS	Mahalanobis space
MTGS	Mahalanobi–Taguchi–Gram–Schmidt process
MTS	Mahalanobis–Taguchi system
μ	Mean
n	Sample size
N_i	Sample size of the ith group
OA	Orthogonal array
PCA	Principal component analysis
QLF	Quality loss function
r	Sum of squares due to the input signal
R^2	Coefficient of determination
RE	Robust engineering
S_β	Sum of squares due to slope
Scaled MD	Scaled Mahalanobis distance
SD	Standard deviation
S_e	Error sum of squares
s_i	Standard deviation of ith characteristic
σ	Standard deviation
σ^2	Standard error
σ_m^2	Variation between abnormals of each class
S/N ratio	Signal-to-noise ratio
SSE	Error sum of squares
S_T	Total sum of squares
T	Threshold
TM	Taguchi Methods
T-value	An indicator of F-ratio
U_i	Gram–Schmidt variables
V_e or σ_e^2	Error variance
VRR	Variability reduction range
$V(X)$	Variance of X
X_i	Original variables
Y_i	Principal components
Z_i	Standardized variables

DEFINITIONS OF MATHEMATICAL AND STATISTICAL TERMS

Adjoint matrix: The adjoint of a square matrix A is obtained by replacing each element of A with its own cofactor and transposing the result.

Chi-square (χ^2) distribution: The distribution of a nonnegative random variable, which is skewed to right. The distribution is specified by the degrees of freedom.

Cofactor: The factor remaining after an element is factored out.

Correlation coefficient (r_{12}): The measure of linear association between the two variables X_1 and X_2. This value lies between -1 and $+1$.

Correlation matrix (C): The matrix that gives correlation coefficients between the variables.

Covariance: A measure of the linear relationship between two variables.

Covariance matrix (C): The matrix that gives covariances between the variables.

Degrees of freedom: The number of independent parameters associated with an entity. These entities could be a matrix experiment, or a factor, or a sum of squares.

Determinant of matrix: A characteristic number associated with a square matrix.

F-distribution: The distribution corresponding to an F-random variable, which is a ratio of two χ^2 random variables. The distribution is specified by degrees of freedom of numerator and denominator.

Inverse matrix: If A and B are two square matrices (of same size) such that $AB = BA = I$ (an identity matrix), then B is called the inverse of A and is denoted by A^{-1}. An inverse matrix exists only for a nonsingular matrix.

Normal distribution: The most commonly used distribution in statistics, also known as a Gaussian distribution. It is a bell-shaped curve and is symmetric about the mean. The distribution is specified by two parameters: mean and standard deviation.

Orthogonal vectors: When the angle between two vectors x and y is 90° or 270°, they are said to be orthogonal. The inner (or dot) product of these vectors is zero.

Projection of a vector: The projection (or shadow) of a vector x on a vector y is given as $(1/L_y^2)\,(x'y)y$, where x' is transpose of vector x, and L_y is the length of vector y.

Singular matrix: A matrix whose determinant is zero; otherwise, the matrix is nonsingular.

Standardized distance: Distance of an observation from the mean in terms of standard deviations.

Standardized variables: Variables obtained after subtracting the mean from the original variables and dividing the subtracted quantity by standard deviation.

Transpose of a matrix: The matrix obtained by interchanging rows and columns of the given matrix.

Variance: Square of the standard deviation.

1
INTRODUCTION

This introductory chapter defines the goal of the book. It discusses the nature of the problem and the proposed approach to its solution and briefly describes the related methods in existing literature. The chapter also serves as a guide to the book by providing a brief description of the various chapters and their relation with each other.

1.1 THE GOAL

Swami Vivekananda, a great Indian saint, emphasized the importance of a goal with the following quote:

Awake, arise and stop not till the goal is reached.

This aptly applies to writing a book, where one has to define a suitable goal and work hard to accomplish it irrespective of the difficulties faced. Therefore, we begin by clearly stating the goal of this book:

The goal of this book is to help decision makers who use multidimensional systems to make robust decisions by recognizing various patterns with an appropriate measurement scale and to use simple procedures to optimize the system.

The decision makers could be doctors, managers, academicians, executives from service organizations, such as banks and insurance companies, or any other type of pattern recognizers. The robust decisions made with the measurement scale help lower the cost of diagnosis and minimize incidents of false diagnosis and diagnosis time.

A *multidimensional system* may be defined as a decision-making (diagnostic or pattern-recognition) system based on the numerous measurements. A multidimensional system could be an inspection system, a medical diagnosis system, a sensor system, a face recognition system, a voice recognition system, or a university admission system. Because multidimensional systems are used in day-to-day life, the proposed methods can find applications in several areas. In this book, case studies and examples related to different types of multidimensional systems are presented.

The main goal can be divided into the following subgoals, which are based on the problems encountered by decision makers in different fields:

- To introduce a measurement scale based on the input characteristics to measure the degree of unhealthiness or abnormality of different conditions
- To quantify functionality of a system with a suitable measure
- To minimize the number of the variables required (in terms of original variables) for an effective diagnosis
- To predict the performance of a multidimensional system under various conditions
- To establish different zones of treatment of a product or patient based on severity and cost so that the decision maker can take appropriate actions
- To identify the direction of abnormal conditions
- To overcome commonly encountered multivariate problems, such as multicollinearity (high correlations) and low correlations

- To demonstrate applicability of proposed methods with actual case studies

This book also compares the proposed methods with classical multivariate methods and artificial neural networks.

1.2   THE NATURE OF A MULTIDIMENSIONAL SYSTEM

To develop a suitable measurement scale for the purpose of diagnosis, it is important to understand the nature of the system in terms of the variables controlling the system. It is also necessary to know the noise conditions under which the diagnosis process is to be performed. Since all the variables may not be necessary for the diagnosis process, it is important to identify the useful set of variables, which is a subset of the original variables. The future diagnosis is performed with the variables in the useful set. While conducting the diagnosis, the decision maker may also be interested in defining a set of actions to be taken for different abnormal conditions.

1.2.1   Description of Multidimensional Systems

A typical multidimensional system used in this research is shown in Figure 1.1. In this figure, X_1, X_2, ..., X_k correspond to k variables (dimensions), which provide information that can be used to make a decision. A correct decision has to be made about the state of the system regardless of noise conditions.

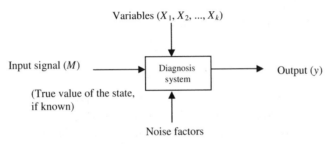

Figure 1.1   Multidimensional diagnosis system.

In Figure 1.1, the input signal M is the true value of the state of the system, if known (for example, a rainfall prediction system where rainfall can be measured and recorded). The noise conditions are the changes in the usage environment, such as conditions in different places and manufacturing variability of the system. The output y should match closely with the input. In future based on the output, a decision has to be made about the state of the system, and accordingly corrective actions have to be taken. In multivariate systems, a correct decision cannot be made by looking at all the variables independently because of the presence of correlations. It is important to take the correlations between the variables into account.

1.2.2 Correlations between the Variables

A measure of linear association between two variables is usually provided by the covariance between them. Covariance between two variables depends on the units of measurement. The correlation coefficient is a standardized version of the covariance and it does not depend on the units of measurements. The patterns of observations in multidimensional systems highly depend on the correlation structure of the variables in the system. One can make wrong decisions about the patterns if each variable is looked at separately without considering the correlation structure.

To construct a multidimensional measurement scale, it is important to have a *distance measure,* which is based on correlations between the variables and by which different patterns could be identified and analyzed with respect to base or reference point. Fortunately, there exists one such measure, called *Mahalanobis distance.* The Mahalanobis distance was introduced in 1936 by P. C. Mahalanobis, a famous Indian statistician and the founder of the Indian Statistical Institute.

1.2.3 Mahalanobis Distance

The Mahalanobis distance (MD) is a generalized distance, which can be considered a single measure of the degree of divergence in the mean values of different characteristics of a population by considering the correlations between the variables. The Mahalanobis distance (Mahalanobis 1936) is a very useful way of determining the similarity of a set of values from an unknown sample to a set of values measured from a collection of known samples. This method has been applied successfully for spectral discrimination in a number of cases. One of the main reasons for using MD is that it is very sensitive to intervariable changes in the reference data. MD is superior to other multidimensional distances, such as Euclidean distance, because it takes distribution of the points (correlations) into account. Traditionally, the Mahalanobis distance is used to classify observations into different groups.

In our approach, the Mahalanobis distance is modified by suitable scaling and is used first to define a base or reference point of the scale with a set of observations from a reference group. The average of the scaled distance in the reference group converges to unity, because of the properties of the scaled distance. Since the reference group has average unit distance, the reference group is also known as a *unit group*. Because the reference group contains scaled Mahalanobis distance, the group is sometimes referred to as *Mahalanobis space* (MS). Mahalanobis space is a database containing the means, standard deviations, and correlation structure of the variables in the reference group. The scaled Mahalanobis distance is also used to measure the distances of unknown observations from the reference point. Defining a reference group or Mahalanobis space and selection of variables for constructing such group depend entirely on the decision maker's discretion.

Since the definition of Mahalanobis space (unit group) is very important in our approach and is based on the information about variables, Figure 1.1 can be modified as shown in Figure 1.2. This figure forms a basis for all the discussions in this book.

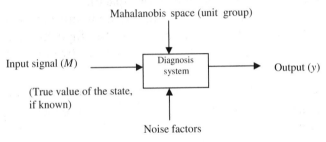

Figure 1.2   Modified multidimensional diagnosis system.

The Mahalanobis distance (and hence the scaled distance) can be computed in two ways: (1) using the inverse of the correlation matrix, and (2) using the Gram–Schmidt orthogonalization process. The advantages of the Gram–Schmidt process are clearly spelled out in this book. A detailed discussion on the Mahalanobis distance and scaled distance is presented in Chapter 2.

1.2.4 Robust Engineering/Taguchi Methods

The concepts of *robust engineering* (RE) are based on the philosophy of Genichi Taguchi, who introduced the concepts after several years of research. Robust engineering systematically evolved starting in the 1950s and aims at providing industries with a cost-effective methodology for enhancing their competitive position in the global market. These concepts are also referred to as *Taguchi Methods.*

In Taguchi Methods, there are two types of quality: (1) customer-driven quality and (2) engineered quality. Customer quality leads to the size of the market segment and includes product features such as color, size, and appearance. The market size becomes bigger as customer quality gets better. Customer quality is addressed during the product planning stage and is extremely important in creating a new market. *Engineered quality* includes the defects, failures, noise, vibrations, pollution, etc. Engineered quality can be measured in terms of deviations from ideal performance (function). While the customer quality defines the market size, the engineered quality helps in winning the market share within the segment. Robust engineering aims at improving the engineered quality. In multivariate applications, *scaled Mahalanobis distance* is similar to engineered quality, because it measures the degree of abnormality (function) of observations from the known reference group (Mahalanobis space).

Since, in this book, the Taguchi Methods are combined with the scaled MD, a brief description of the Taguchi Methods is provided here. Taguchi Methods are based on five principles.

1. Measurement of function using energy transformation
2. Taking advantage of interactions between control and noise factors
3. Use of orthogonal arrays and signal-to-noise ratios

4. Two-step optimization
5. Tolerance design using quality loss function and on-line quality engineering

Taguchi (1987), Phadke (1989), and Park (1996) provide a detailed discussion on the principles of TM. These principles are extremely useful and cost effective and have been successfully applied in many engineering applications to improve the performance of a product/process. A brief illustration of these principles is given below:

1. *Measurement of Function Using Energy Transformation*
   The most important aspect of Taguchi Methods (TM) is to find a suitable function (called an ideal function) that governs the system. Taguchi Methods state that "to improve quality, don't measure quality." It is important, instead, to measure the functionality of the system to improve the product performance (quality).

2. *Take Advantage of Interactions between Control and Noise Factors*
   In TM, we are not interested in measuring the interaction between the control factors. We are interested in the interaction between the control and noise factors, since the objective is to make design robust against the noise factors.

3. *Use of Orthogonal Arrays (OAs) and Signal-to-Noise (S/N) Ratios*
   OAs are used to minimize the number of runs (or combinations) needed for the experiment. Many people are of the opinion that the application of OA is TM, but the application of OAs is only a part of TM. S/N ratios are used as a measure of the functionality of the system. S/N ratios capture the magnitude of real effects (signals) after making some adjustment to uncontrollable variation (noise).

4. *Two-Step Optimization*
   After conducting the experiment, the factor–level combination for the optimal design is selected with the help of two-step optimization. The first step is to minimize the variability (maximize S/N ratios). In the second step, the sensitivity (mean) is adjusted to the desired level since it is easier to adjust the mean after minimizing the variability.

5. *Tolerance Design Using Quality Loss Function and On-Line Quality Engineering*
While the first four principles are related to parameter design, the fifth principle is related to the tolerance design and on-line quality engineering (QE). Having determined the best settings using parameter design, the tolerancing is done with the help of quality loss function. If the performance deviates from the target, a loss is associated with the deviation, known as a loss to the society. This loss is proportional to the square of the deviation. It is recommended that safety factors be designed using this approach. On-line QE is used to monitor the process performance and detect the changes in the process.

1.3 MULTIVARIATE DIAGNOSIS—THE STATE OF THE ART

A significant body of literature exists on the concepts of multivariate diagnosis that are being used in various multidimensional systems. So, where is the need for another book on multivariate diagnosis? This section addresses this question by briefly reviewing some of the major tools used to analyze multidimensional systems and identifying the limitations of the available tools.

1.3.1 Principal Component Analysis

Principal component analysis (PCA) is used for explaining the variance–covariance structure through a fewer linear combinations of original variables. The objectives of PCA are (1) data reduction and (2) data interpretation. Although p components are required to reproduce the total system variability, often much of this variability can be explained by k ($k < p$) principal components. The principal components are particular linear combinations of the p random variables, X_1, X_2, X_3, ..., X_p. These linear combinations represent the selection of a new coordinate system obtained by rotating the original system with maximum variability to provide a simpler description of the covariance structure. Johnson and Wichern (1992) provide a clear discussion on PCA. To calculate the principal components, we need all original variables. Hence, PCA is not helpful in reducing the dimensionality in terms of original variables.

1.3.2 Discrimination and Classification Method

The discrimination and classification method is a multivariate technique concerned with separating the distinct sets of objects or observations and allocating new objects or observations to previously defined groups. Discriminant analysis is carried out using a discriminant function or Mahalanobis distance. The discriminant function is a part of Mahalanobis distance, which provides another rule for classification.

When there are k populations, in the first stage discriminant functions are developed for all the groups. It is assumed that all the groups have same covariance matrix. In the second stage, the classification of a new observation is done based on the following rule:

- Assign a new observation X to the group whose mean is closest to this observation (minimum Mahalanobis distance), or
- Assign a new observation X to the group that has the largest discriminant function.

The classification is done in such a way that the expected cost of misclassification (ECM) is minimized. ECM computations are based on the cost of misclassification and prior probabilities. Therefore, the method of discrimination and classification is probabilistic in nature. A good discussion on this method is provided in Johnson and Wichern (1992). Discrimination and classification methods have certain limitations. In these methods, the emphasis is given to classification of an observation in a group. These methods are less helpful for accurately measuring the level of severity of an abnormality in order to take appropriate corrective actions.

1.3.3 Stepwise Regression

Stepwise regression is widely used for selection of a useful subset of variables in multivariate applications. The procedure iteratively constructs a sequence of regression models by adding and removing variables at each step. Based on a specified value of the F-random variable, addition and deletion of the variables in the model is carried out. This method requires several iterations, if the number of variables is high. The method of stepwise regression has been criticized because it does not guarantee the best subset regression model. A discussion on stepwise regression is given in Montgomery and Peck (1982).

1.3.4 Test of Additional Information (Rao's Test)

The test of additional information, also known as Rao's test, is used to identify a set of useful variables. In Rao's test, which uses Fischer's linear discrimination function, subsets are tested for significance by computing an F-statistic. A high F-ratio indicates that the subset of q variables provides additional information on the discriminant analysis. If the F-ratio is not high, then we can discard the subset of variables. For this test procedure please refer to Rao (1973). In this method, selection of the subset of variables is carried out based on prior knowledge about variables or expert opinion. Moreover, testing the significance of variables using an F-test may not be adequate to decide the important variables. This fact is illustrated in Section 9.4.

1.3.5 Multiple Regression

In multiple regression (MR), the characteristic y (dependent variable) is estimated based on the p independent variables $X_1, X_2, ..., X_p$. Based on the value of y, a decision can be made regarding the classification of an observation X, which consists of $X_1, X_2, ..., X_p$. A discussion of multiple regression is provided in Montgomery and Peck (1982). MR models are developed using least-squares estimates and are based on certain assumptions about the error term. MR models may become complex if the number of variables is high.

1.3.6 Multivariate Process Control Charts

Multivariate process control charts are an extension of univariate control charts, where more than one variable is monitored and controlled over a period of time. Several types of these charts, such as multivariate Shewhart charts and multivariate Cusum charts, are being used. The purpose of these charts is to monitor and control the multivariate conditions. These charts operate just like univariate charts, in which corrective actions are taken whenever the process is out of the control limits. An extensive literature is available on multivariate control charts. In multivariate charts, generally, the variables $X_1, X_2, ..., X_p$ are assumed to follow a p-dimensional normal distribution and therefore the control limits are probabilistic.

1.3.7 Artificial Neural Networks

Artificial neural networks (ANN) are used for pattern recognition, learning, classification, generalization, and interpretation of noisy inputs. A structure (network) is composed of interconnected units (artificial neurons). Each unit has an input/output (I/O) characteristic and implements a local computation or function. The output of any unit is determined by its I/O characteristic and its interconnection to other units and (possibly) external inputs. ANN constitutes not one network, but a diverse family of networks. ANN and the proposed methods in this book do not require any probabilistic assumptions, but ANN have certain limitations:

- The data (patterns) are to be randomized for training the network.
- ANN will not provide a relationship between input and output.
- The dimensionality reduction cannot be easily done.
- The degree of abnormality cannot be measured on a scale.

An important goal of this book is to overcome the limitations of existing multivariate/pattern recognition methods. In the proposed methods, the problem of multivariate diagnosis is viewed with an entirely different perspective. In Chapter 10, a detailed discussion of classical multivariate methods and neural networks is provided. These methods are also compared with proposed methods by using suitable examples.

1.4 APPROACH

In this section, the overall approach to meeting the goal of the book is discussed. The approach is developed by considering the nature of problems faced by decision makers while dealing with multidimensional systems. In proposed methods the scaled Mahalanobis distance is used to develop a measurement scale for the multidimensional systems, and the principles of Taguchi Methods (TM) are used to optimize the system and predict its performance. Therefore, this methodology is referred to as the *Mahalanobis–Taguchi System (MTS)*. In MTS, scaled MDs are computed using the inverse of the correlation matrices. If the scaled MDs are computed using the Gram–Schmidt orthonormalization process, then such a method is referred to as the *Mahalanobis–Taguchi–Gram–Schmidt (MTGS) method*. MTS and MTGS methods are discussed in Chapters 2 and 3.

1.4.1 Classification versus Measurement

In classical methods, such as discrimination and classification methods (and sometimes in multiple regression), the objective is to classify the observations into different groups (populations). On the contrary, the main objective of the methods proposed in this book is to provide a measurement scale to measure the degree of abnormalities on a continuous scale. This will help determine the appropriate actions to take based on the degree of abnormalities.

1.4.2 Normals versus Abnormals

In classical methods, both the normal group and abnormal group are considered separate populations. The classification is based on the distances of an observation from the means of these populations. In the MTS/MTGS methods, there are no populations. We need only a group of observations called a "normal" or "healthy" group to obtain correlation structure and to define the reference point to the measurement scale. Selection of this group is entirely at the discretion of decision maker. In the MTS/MTGS methods, every abnormal condition (or a condition outside "healthy" group) is considered unique, since the occurrence of such a condition is different. The degree of abnormality is measured in reference to the normal group. In this context, it is worthwhile to note Tolstoy's quote in *Anna Karenina:*

All happy families look alike. Every unhappy family is unhappy after its own fashion.

1.4.3 Probabilistic versus Data Analytic

In classical multivariate methods, probability-based inference is used for analyzing multivariate systems. For example, in discrimination and classification methods, the cost of misclassification and probabilities are used for classification and the stepwise regression models are probabilistic. On the contrary, in the MTS/ MTGS methods, the measures and procedures used are data analytic. The quadratic loss function concept is used for determining the value of a threshold. The loss function approach minimizes the total cost by considering various cost elements. This method of finding threshold is not probabilistic—it uses the measures of descriptive statistics. Based on the threshold, the multivariate systems can be monitored and appropriate actions can be taken accordingly.

1.4.4 Dimensionality Reduction

In multivariate systems, dimensionality reduction is still a challenge. Principal component analysis is used to reduce the dimensionality of the systems by computing the principal components. However, to calculate one component, we need the entire original set of variables. In other words, this technique does not provide a methodology for dimensionality reduction in terms of original variables. Methods like test of additional information, though intended for dimensionality reduction in terms of original variables, require prior knowledge about the variables. They also depend on the F-ratio, which may not be sufficient to identify the variables of importance. Stepwise regression also depends on an F-ratio. This technique requires several iterations if the number of variables is large.

We propose the use of S/N ratios to identify the variables of importance. Based on S/N ratios, we can directly obtain the useful set of original variables.

1.5 REFINING THE SOLUTION STRATEGY

Given the preference to a data analytic and simple methods for multivariate diagnosis, the top-level solution strategy integrates Mahalanobis distance and the principles of Taguchi Methods. Having chosen a top-level solution strategy, the subrequirements can be defined as shown in the Figure 1.3. In addition to having an effective diagnostic procedure, it may be necessary to make robust decisions under the influence of noise factors, to identify the direction of abnormals and to overcome effects of multicollinearity and small correlations. The solution strategy developed takes these aspects into account.

Much of this book is concerned with expanding the solution strategy shown in Figure 1.3. This book will provide a set of equations, concepts, and procedures that help decision makers make effective decisions while diagnosing multidimensional systems. The book aims at continually measuring multidimensional systems with the help of simple measures and procedures.

1.6 GUIDE TO THIS BOOK

Most of the chapters in this book are developed based on Chapter 2, which provides an introduction to MTS/MTGS methods. To get an idea about these methods and how they differ from existing methods and for actual cases, readers can browse through Chapters 10 and 11 using Chapter 2 as a basis. The following are brief chapter summaries of the book:

Chapter 1 is an introductory chapter that defines the goal of the book, discusses the nature of the problem, and proposes an approach to its solution. The chapter briefly discusses the related methods in existing literature.

Chapter 2 introduces the MTS and MTGS methods. A comparison is made between these two methods with the help of medical diagnosis data.

Chapter 3 discusses the advantages and limitations of MTS and MTGS methods. The examples provided in this chapter include a medical diagnosis system, a graduate admission sys-

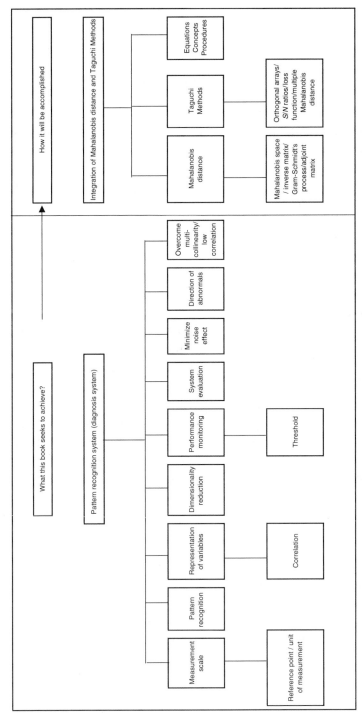

Figure 1.3 Solution strategy.

tem, and the data set provided by American Supplier Institute (ASI).

Chapter 4 introduces the role of orthogonal arrays (OAs) and signal-to-noise (*S/N*) ratios in multivariate diagnosis. The chapter shows how dimensionality reduction can be performed with the help of OAs and *S/N* ratios. The chapter also discusses the advantages of using *S/N* ratios in multivariate diagnosis.

Chapter 5 explains the applicability of MTS/MTGS methods for categorical data. This feature is explained using a sales promotion case study.

Chapter 6 describes different ways of treating noise factors in MTS/MTGS methods. Noise conditions are changes in usage environment, such as conditions in different places and manufacturing variability of the systems.

Chapter 7 describes the role of quality loss function in determining thresholds for MTS/MTGS methods. These methods cannot use the prior probabilities and expected cost of misclassification because they are data analytic.

Chapter 8 discusses the standard error of the measurement scale developed using MTS/MTGS methods.

Chapter 9 focuses on advanced topics related to MTS/MTGS methods. These topics are included to provide suitable procedures to overcome problems due to multicollinearity and small correlations, to select subsets for complex cases, and to select a good Mahalanobis space from historical data.

Chapter 10 compares MTS/MTGS with classical multivariate statistical methods and artificial neural networks, using a medical diagnosis example.

Chapter 11 describes the applicability of this methodology to the case studies in different areas, representing both American and Japanese industry.

Chapter 12 concludes by highlighting the important points of proposed methods, their scientific contributions, limitations, and directions for future research.

2
MTS AND MTGS

This chapter introduces the methodologies of the Mahalanobis–Taguchi system (MTS) and the Mahalanobis–Taguchi–Gram–Schmidt process (MTGS). Basic steps in these methods are described with the help of medical diagnosis data. We begin with a detailed discussion of Mahalanobis distance.

2.1   A DISCUSSION Of MAHALANOBIS DISTANCE

Mahalanobis distance measures distances in multidimensional spaces by taking correlations into account. MD is very sensitive to the correlation structure of the reference group. In classical methods, MD is used to find the "nearness" of an unknown point from the mean point of a group(s). The observation is classified into a group from whose center it has least distance. There are other multivariate measurement techniques, such as the Euclidean distance (ED). Euclidean distance also gives the distance of the "unknown" point from the group mean point, but there are two disadvantages of this technique: (1) Euclidean distance does not give a statistical measurement of how well the unknown matches the reference set, and (2) it measures only a relative distance from the mean point in the group and does not take into account the distribution of the points in the group.

Figure 2.1 shows a comparison between MD and ED. Let us assume that there is reference set consisting of observations on variables X_1 and X_2. In this figure, the elliptical shape refers to the Mahalanobis boundary and the circular shape refers to Euclidean boundary of the reference set. Let us assume that two unknown sample points A and B have been added to the system, as shown in the figure. By the Euclidean distance method, sample B is likely to be classified as belonging to the group containing sample A because the relative distance from the center of circular boundary of these points is same. However, sample A clearly lies along the elongated axis of the reference group points, indicating that sample A is behaving much more like the reference group and is not similar to sample B. This will be clear if we compute MD instead of ED, because the ED method does not take into account the correlations between variables (distribution of points). Therefore, ED is less helpful than MD in multivariate analysis. The computational procedure for MD is discussed in the subsequent sections of this chapter.

In MTS/MTGS methods, the Mahalanobis distance is modified by suitable scaling. It is used to define a base or reference point of the scale and to measure the distances of unknown observations from the reference point.

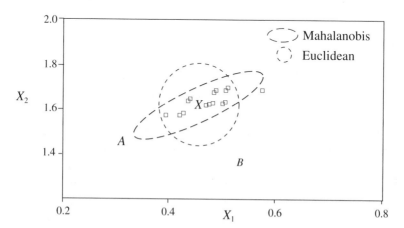

Figure 2.1   Mahalanobis distance and Euclidean distance.

2.2 OBJECTIVES OF MTS AND MTGS

One of the primary objectives of MTS/MTGS methods is to introduce a scale based on all input characteristics to measure the degree of abnormality. To construct such a scale, Mahalanobis distance (MD) is suitably scaled by dividing the original distance by the number of variables k. This scaled MD is similar to the mean square deviation (MSD).

In MTS/MTGS, the first step is to define a "reference" or "normal" or "healthy" group, called Mahalanobis space (MS). In medical diagnosis, MS is constructed only for the people who are healthy, and in a manufacturing inspection system the MS is constructed for high-quality products. Thus, MS is a database for the normal group consisting of the following quantities:

m_i = mean vector of the healthy group
s_i = standard deviation vector of the healthy group
C = correlation matrix of the healthy group

It can be shown (with standardized values) that MS has the zero point as its mean with an average MD equal to unity. This proof is given in Section 2.2.3. Because the average MD of MS is unity, MS is also called the unit space. The zero point and the unit distance are properties of MS that serve as the reference point and base of the measurement scale. Defining the normal group (or MS) is the choice of the specialist conducting the diagnosis. This scale should also be good for measuring conditions outside the MS. To validate the scale, different known conditions with different degrees of severity must be checked. If the scale is good, the MDs corresponding to these conditions should match with judgment. As mentioned before, in MTS/MTGS, the abnormal conditions (conditions outside MS) are not considered a separate group because the occurrence of each abnormality or a condition outside MS is unique. For example, a patient may be outside MS because of high blood pressure or because of high cholesterol content. For this reason, the same correlation matrix (of the MS) is used to compute the Mahalanobis distances of the conditions outside MS. In the next phase of MTS/MTGS, OAs and S/N

ratios are used to choose the variables of importance. There are different kinds of S/N ratios, depending on the prior knowledge about the severity of the conditions. A detailed discussion on OAs and S/N ratios is provided in the Chapter 4.

2.2.1   Mahalanobis Distance (Inverse Matrix Method)

The Mahalanobis distance is a squared distance (also denoted as D^2) calculated for the jth observation in a sample of size n with k variables using the following formula:

$$\text{MD}_j = D_j^2 = Z'_{ij} \, C^{-1} \, Z_{ij} \qquad (2.1)$$

where

$$
\begin{aligned}
j &= 1 \text{ to } n \\
Z_{ij} &= (z_{1j}, z_{2j}, \dots, z_{kj}) \\
&= \text{standardized vector obtained by standardized values of } X_{ij} \\
&\quad (i = 1, \dots, k) \\
Z_{ij} &= (X_{ij} - m_i)/s_i \\
X_{ij} &= \text{value of the } i\text{th characteristic in the } j\text{th observation} \\
m_i &= \text{mean of the } i\text{th characteristic} \\
s_i &= \text{standard deviation (SD) of the } i\text{th characteristic} \\
' &= \text{transpose of the vector} \\
C^{-1} &= \text{inverse of the correlation matrix}
\end{aligned}
$$

In MTS, the MD obtained from (2.1) is scaled by dividing with the number of variables k. Therefore, the equation for calculating scaled MD becomes

$$\text{MD}_j = \frac{1}{k} Z'_{ij} \, C^{-1} \, Z_{ij} \qquad (2.2)$$

MD can also be computed using the Gram–Schmidt orthogonalization process (GSP), which is described in the following section. Johnson and Wichern (1992) provide a discussion on GSP.

2.2.2   Gram–Schmidt Orthgonalization Process

Given linearly independent vectors $Z_1, Z_2, ..., Z_k$, there exist mutually perpendicular vectors $U_1, U_2, ..., U_k$ with the same linear span (Figure 2.2). The Gram–Schmidt vectors are constructed sequentially by setting

$$U_1 = Z_1$$

$$U_2 = Z_2 - \frac{Z_2' U_1}{U_1' U_1} U_1$$

$$\vdots$$

$$U_k = Z_k - \frac{Z_k' U_1}{U_1' U_1} U_1 - \cdots - \frac{Z_k' U_{k-1}}{U_{k-1}' U_{k-1}} U_{k-1}$$

where a prime denotes the transpose of a vector. When MD is calculated using GSP, standardized values of the variables are used. Therefore, in the above set of equations Z_1, Z_2, ..., Z_k correspond to standardized values. It is clear that the transformation process largely depends on the first variable.

Calculation of MD Using the Gram–Schmidt Process

Suppose that we have a sample of size n and that each sample contains observations on k variables. After standardizing the variables, we will have a set of standardized vectors:

$$Z_1 = (z_{11}, z_{12}, ..., z_{1n})$$

$$Z_2 = (z_{21}, z_{22}, ..., z_{2n})$$

$$\vdots$$

$$Z_k = (z_{k1}, z_{k2}, ..., z_{kn})$$

After performing GSP, we have orthogonal vectors:

Figure 2.2   Gram–Schmidt process.

$$U_1 = (u_{11}, u_{12},...,u_{1n})$$

$$U_2 = (u_{21}, u_{22},...,u_{2n})$$

$$\vdots$$

$$U_k = (u_{k1}, u_{k2},...,u_{kn})$$

It easily follows that the mean of vectors $U_1, U_2,...,U_k$ is zero. Let $s_1, s_2,...,s_k$ be standard deviations of $U_1, U_2,...,U_k$, respectively. Since we have a sample of size n, there will be n MDs. The MD corresponding to the jth observation of the sample is computed using the following equation:

$$MD_j = \frac{1}{k}\left(\frac{u_{1j}^2}{s_1^2} + \frac{u_{2j}^2}{s_2^2} + \cdots + \frac{u_{kj}^2}{s_k^2}\right) \tag{2.3}$$

where $j = 1,...,n$. The values of MD obtained from (2.2) and (2.3) are exactly the same, which is proved in the following section.

2.2.3   Proof That Equations 2.2 and 2.3 Are the Same

We know that the Gram–Schmidt vectors U_1, U_2, ...,U_k are mutually perpendicular vectors. Let $s_1, s_2,...,s_k$ be SDs of U_1, U_2, ...,U_k, respectively. Since these vectors are obtained from the standardized vectors $Z_1, Z_2,...,Z_k$, the average of these vectors is equal to zero. Using Equation 2.2, we have

$$MD_j = \left(\frac{1}{k}\right)(U_{ij}'\, C^{-1}\, U_{ij})$$

where

$$C = \begin{bmatrix} s_1^2 & 0 & 0 & \cdots & 0 \\ 0 & s_2^2 & 0 & \cdots & 0 \\ \multicolumn{5}{c}{\dotfill} \\ 0 & 0 & 0 & \cdots & s_k^2 \end{bmatrix}$$

Since this is a diagonal matrix,

$$C^{-1} = \begin{bmatrix} 1/s_1^2 & 0 & 0 & \cdots & 0 \\ 0 & 1/s_2^2 & 0 & \cdots & 0 \\ \multicolumn{5}{c}{\cdots\cdots\cdots\cdots\cdots\cdots\cdots\cdots\cdots} \\ 0 & 0 & 0 & \cdots & 1/s_k^2 \end{bmatrix}$$

and $U_{ij}' = (u_{1j}, u_{2j},...,u_{kj})$, where $j = 1,...,n$. Therefore, after performing the matrix multiplication,

$$\text{MD}_j = \frac{1}{k}\left(\frac{u_{1j}^2}{s_1^2} + \frac{u_{2j}^2}{s_2^2} + \cdots + \frac{u_{kj}^2}{s_k^2}\right)$$

which is Equation 2.3. Hence, the result. $\square$

2.2.4 Calculation of Mean of the Mahalanobis Space

It has been proved that MD (without scaling) follows a χ^2 distribution with k degrees of freedom, when the sample size n is large and all characteristics follow the normal distribution. This proof can be obtained from Johnson and Wichern (1992). We know that a χ^2 statistic with k degrees of freedom has a mean equal to k. Hence, scaled MD has a mean of 1.0.

In fact, the assumptions of distribution of input variables are not necessary for calculating MDs in MS. We prove this in the following steps.

Proof That MD Can Be Obtained without Assumption of Distribution of Variables

We start by considering the univariate case. Let X_j ($j = 1,...,n$) be a variable with a mean μ and SD σ. The quantity $(X_j - \mu)/\sigma$, known as the standardized variable, measures the distance of X_j from μ in terms of SD units. The quantity $(X_j - \mu)^2/\sigma^2$ measures the squared distance of X_j from μ in terms of SD units. The measurement scale for such quantities will start from the zero and all the measured distances are positive. It can be easily shown that the expected value of $(X_j - \mu)^2/\sigma^2$ is equal to 1.0. This can be mathematically written as

$$E\left[\frac{(X_j - \mu)^2}{\sigma^2}\right] = 1.0 \tag{2.4}$$

Now, the quantity $(X_j - \mu)^2/\sigma^2$ can be written as $(X_j - \mu)(1/\sigma^2)$ $(X_j - \mu)$ or

Squared standardized distance $= (X_j - \mu) \text{ (Variance)}^{-1} (X_j - \mu)$

$$\tag{2.5}$$

From Equation 2.2, we have $\text{MD}_j = (1/k)Z_{ij}'C^{-1} Z_{ij}$, where $i = 1, ..., k$ ($k =$ number of characteristics/variables). Equations 2.5 and 2.2 are similar because Equation 2.5 measures squared distance in the univariate case and Equation 2.2 measures squared distance in the multivariate case. In the multivariate case, since we have to take correlations into account, the correlation matrix replaces the variance term.

MD can also be represented in terms of the Gram–Schmidt variables using Equation 2.3, which is given as

$$\text{MD}_j = \left(\frac{1}{k}\right)\left(\frac{u_{ij}^2}{s_1^2} + \frac{u_{2j}^2}{s_2^2} + \cdots + \frac{u_{kj}^2}{s_k^2}\right)$$

In this equation, u's are the Gram–Schmidt variables and $s_1, s_2,...,s_k$ are the SDs of these variables. Since the Gram–Schmidt process is carried out with the help of standardized variables, the means of $u_{1j}, u_{2j},...,u_{kj}$ are zeros. The squared terms of Equation 2.3 correspond to the squared distances of the Gram–Schmidt variables from the respective means in terms of SD units. The expected value of squared distances (inside the parentheses) in Equation 2.3 is equal to 1.0 (from Equation 2.4).

Now,

$$E(\text{MD}_j) = E\left[\frac{1}{k}\left(\frac{u_{1j}^2}{s_1^2} + \frac{u_{2j}^2}{s_2^2} + \cdots + \frac{u_{kj}^2}{s_k^2}\right)\right]$$

or

$$E(\text{MD}_j) = \left(\frac{1}{k}\right)\left[E\left(\frac{u_{1j}^{2}}{s_1^{2}} + \frac{u_{2j}^{2}}{s_2^{2}} + \cdots + \frac{u_{kj}^{2}}{s_k^{2}}\right)\right]$$

Since Gram Schmidt's vectors are orthogonal, we can write

$$E(\text{MD}_j) = \frac{1}{k}\left[E\left(\frac{u_{1j}^{2}}{s_1^{2}}\right) + E\left(\frac{u_{2j}^{2}}{s_2^{2}}\right) + \cdots + E\left(\frac{u_{kj}^{2}}{s_k^{2}}\right)\right]$$

Since expected values on the right-hand side are equal to 1.0, we have

$$E(\text{MD}_j) = \left(\frac{1}{k}\right)\frac{[1 + 1 + \cdots + 1]}{K \text{ times}}$$

$$E(\text{MD}_j) = \frac{1}{k}\,[k] = 1$$

Therefore, scaled Mahalanobis distance has properties of the quantity $(X_i - \mu)^2/\sigma^2$ because it measures distances from the zero point and has an expected value equal to unity. $\square$

With this proof, we can say that we do not require the assumption of any distribution to define the zero point and unit distance and hence it can be generalized for any number of variables irrespective of their distributions (including categorical data). MS is centered at zero point because the original variables are converted into standardized variables. We cannot define the zero point and unit distance if standardized variables are not used and scaling is not done.

The advantages of defining MS with scaled MDs are that

- The definition of MS can be generalized to any number of variables.
- The average value of MDs in MS is always 1.0.

Note: From this point on, for simplicity, the scaled MD is denoted as MD.

2.3 STEPS IN MTS

Stage I: Construction of a Measurement Scale with Mahalanobis Space (Unit Space) as the Reference

- Define the variables that determine the healthiness of a condition. For example, in a medical diagnosis application, the doctor has to consider the variables of all diseases to define a healthy group. In general, for pattern recognition applications, the term "healthiness" must be defined with respect to "reference pattern."
- Collect the data on all the variables from the healthy group.
- Compute the standardized values of the variables of the healthy group.
- Compute MDs of all observations using the inverse of the correlation matrix. With these MDs, we can define the zero point and the unit distance.
- Use the zero point and the unit distance as the reference point or base for the measurement scale.

Stage II: Validation of the Measurement Scale

- Identify the abnormal conditions. In medical diagnosis applications, the abnormal conditions refer to the patients having different kinds of diseases. In fact, to validate the scale, we may choose any condition outside MS.
- Compute the MDs corresponding to these abnormal conditions to validate the scale. The variables in the abnormal conditions are normalized by using the mean and SDs of the corresponding variables in the healthy group. The correlation matrix corresponding to the healthy group is used to compute the MDs of abnormal conditions.
- If the scale is good, the MDs corresponding to the abnormal conditions should have higher values. In this way the scale is validated. In other words, the MDs of conditions outside MS must match with judgment.

Stage III: Identify the Useful Variables (Developing Stage)

- Find out the useful set of variables using orthogonal arrays (OAs) and *S/N* ratios. The *S/N* ratio, obtained from the abnormal MDs, is used as the response for each combination of OA. The useful set of variables is obtained by evaluating the "gain" in the *S/N* ratio.

Stage IV: Future Diagnosis with Useful Variables

- Monitor the conditions using the scale, which is developed with the help of the useful set of variables. Based on the values of MDs, appropriate corrective actions can be taken. The decision to take the necessary actions depends on the value of the threshold.

These stages remain the same even if the GSP process (MTGS method) is used for diagnosis. For the purpose of better understanding, the steps in MTGS are given in the following section.

2.4 STEPS IN MTGS

Stage I: Construction of a Measurement Scale with Mahalanobis Space (Unit Space) as the Reference

- Define the variables that determine the healthiness of a condition.
- Collect the data on all variables from the healthy group.
- Compute the standardized values of the variables of the healthy group.
- Compute MDs of all observations using the Gram–Schmidt orthogonalization process. With these MDs we can define the zero point and the unit distance.
- Use the zero point and the unit distance as the reference point or base for the measurement scale.

Stage II: Validation of the Measurement Scale

- Identify the abnormal conditions. In medical diagnosis applications, the abnormal conditions refer to the patients having different kinds of diseases.
- Compute the MDs corresponding to these abnormal conditions to ensure the accuracy of the scale. We may also choose any condition outside MS to validate the scale. The variables in the abnormal conditions are normalized by using the mean and SDs of the corresponding variables in the healthy group. The Gram–Schmidt coefficients corresponding to the healthy group are used to compute the MDs of abnormal conditions.
- If the scale is good, the MDs corresponding to the abnormal conditions should have higher values or MDs of outside conditions must match with judgment. In this way the scale is validated.

Stage III: Identify the Useful Variables (Developing Stage)

- Find out the useful set of variables by using S/N ratios. In MTGS, S/N ratios can be computed for all the variables directly from the orthonormal vectors, if the effects of partial correlations* are not significant and if we can define a specific order for these variables, since the Gram–Schmidt process depends on the first variable.
- However, if the effects of partial correlations are significant, then OAs are required to find a set of useful variables. It is recommended that OAs always be used, because one need not be concerned with the order of the variables or partial correlations

Stage IV: Future Diagnosis with Useful Variables

- Monitor the conditions using the scale, which is developed with the help of the useful set of variables. Based on the values of MDs, appropriate corrective actions can be taken.

---

*Please refer to Chapter 3 for a discussion on partial correlations.

The decision to take the necessary actions depends on the value of the threshold.

2.5   DISCUSSION OF MEDICAL DIAGNOSIS DATA: USE OF MTGS AND MTS METHODS

To compare MTS and MTGS methods, we will use a Japanese doctor's (Dr. Kanetaka's) data on liver disease testing. The data contain observations of a healthy group as well as of the abnormal conditions. The 17 variables considered for the purpose of diagnosis are as shown in Table 2.1. The healthy group (MS) is constructed based on observations on 200 people, who do not have any health problems. There are 17 abnormal conditions. The comparison is made using the visual basic programs written exclusively for this purpose. The data sets are the inputs to the software programs. The results (in terms of MDs) obtained by the two software programs in the case of the healthy group and abnormals are presented in Tables 2.2 and 2.3 respectively.

From these tables, it is clear that the results of the two methods are the same. In both methods, the average MD of the healthy

TABLE 2.1   Variables in Medical Diagnosis Data

S.No.	Variables	Notation	Notation for Analysis
1	Age		X_1
2	Sex		X_2
3	Total protein in blood	TP	X_3
4	Albumin in blood	Alb	X_4
5	Cholinesterase	ChE	X_5
6	Glutamate O transaminase	GOT	X_6
7	Glutamate P transaminase	GPT	X_7
8	Lactate dehydrogenase	LHD	X_8
9	Alkanline phosphatase	Alp	X_9
10	r-Glutamyl transpeptidase	r-GPT	X_{10}
11	Leucine aminopeptidase	LAP	X_{11}
12	Total cholesterol	TCh	X_{12}
13	Triglyceride	TG	X_{13}
14	Phospholopid	PL	X_{14}
15	Creatinine	Cr	X_{15}
16	Blood urea nitrogen	BUN	X_{16}
17	Uric acid	UA	X_{17}

TABLE 2.2   Healthy Group MDs Obtained by MTS and MTGS Methods

	1	2	3	4	...	197	198	199	200	Average
MTS	0.378	0.431	0.404	0.500	...	1.754	1.777	1.756	2.358	0.995
MTGS	0.378	0.431	0.403	0.500	...	1.754	1.777	1.756	2.358	0.995

group converges to 1.0. Higher values of MDs for abnormals validate the accuracy of the measurement scale.

Minimization of the Number of Variables Using the S/N Ratio

In both methods, the S/N ratio (measure of accuracy of the measurement scale) is used to identify the useful variables. Since in this application the true levels of severity of abnormals are not known, S/N ratios are calculated on the basis of a "larger-the-better" criterion. Larger-the-better-type S/N ratios are used, because MDs of abnormals must be as high as possible. The procedures, used in both methods, for calculating S/N ratios (and hence useful variables) are presented below.

MTGS Method

Let there be t abnormal conditions. Let u_{i1}, u_{i2},...,u_{it} be the elements of the ith Gram–Schmidt variable (here $i = 1$ to k). Let s_i be the SD of this variable. Then the S/N ratio η_i (for the larger-the-better criterion) corresponding to this variable is

$$\eta_i = -10 \log \left\{ \frac{1}{t} \sum_{j=1}^{t} \left[\frac{1}{(U_{ij}/s_i)^2} \right] \right\} \qquad (2.6)$$

In the MTGS method, the S/N ratio is calculated with the help of Equation 2.6 for the larger-the-better criterion. We can also cal-

TABLE 2.3   Abnormal MDs Obtained by MTS and MTGS Methods

	1	2	3	4	...	14	15	16	17	Average
MTS	7.73	8.42	10.29	7.21	...	43.04	78.64	97.27	135.71	30.39
MTGS	7.73	8.42	10.29	7.20	...	43.04	78.64	97.27	135.70	30.39

culate "dynamic" *S/N* ratios if the required information is available. Note that Equation 2.6 gives *S/N* ratios for orthonormal variables. Use of this equation is recommended if partial correlations are not significant and there exists a definite order of the variables.

This chapter provides only an introduction to MTS/MTGS methods; detailed discussions of *S/N* ratios and other related topics are presented in Chapter 4.

Sample values of the *S/N* ratios (in dB) along with the GSP vectors are given in Table 2.4. In Table 2.4, GS vectors for only 8 abnormal conditions are shown. Actually, there are 17 conditions of abnormality. In Table 2.4, $U_1,U_2,...,U_{17}$ are the Gram–Schmidt variables corresponding to original variables $X_1,X_2,...,X_{17}$, respectively. From Table 2.4, it is clear that the variables $U_2,U_5,U_6,U_7,U_{10},U_{12},U_{13},U_{14}$, and U_{15} have higher *S/N* ratios as compared to others. Therefore, these variables are considered useful variables, which means they are sufficient to construct the mea-

TABLE 2.4 GSP Vectors for Abnormal Data and *S/N* Ratios (Sample Values)

	1	2	3	4	5	6	7	8	*S/N* Ratio
U_1	1.19	1.58	0.71	0.80	0.71	0.61	0.61	1.48	−6.74
U_2	−1.01	−0.89	−1.15	0.97	0.94	0.91	0.91	1.17	0.87
U_3	1.86	−1.83	−1.72	3.09	0.65	−0.07	−0.07	0.17	−13.89
U_4	−0.75	−1.10	−0.84	−1.96	−1.17	0.17	0.17	−1.34	−5.31
U_5	−3.29	−3.82	−3.47	−3.47	−4.10	−3.48	−3.48	−4.08	14.15
U_6	1.59	3.27	0.90	1.27	−0.32	3.99	3.99	0.44	−1.91
U_7	0.47	1.34	−0.30	0.80	0.29	5.45	5.45	1.32	0.58
U_8	3.25	0.72	3.06	2.78	−1.12	2.05	2.05	1.98	−5.24
U_9	0.09	2.72	0.32	−0.14	2.41	1.05	1.05	−0.89	−10.30
U_{10}	−0.24	1.87	0.67	0.44	4.84	−1.27	−1.27	0.57	1.18
U_{11}	−0.49	0.25	0.09	−1.31	−0.29	−0.91	−0.91	−0.20	−18.91
U_{12}	4.58	0.76	0.78	2.96	−0.49	2.85	2.85	3.68	3.75
U_{13}	5.29	0.65	4.17	5.44	1.71	0.35	0.35	8.76	−2.95
U_{14}	0.55	2.03	1.73	0.55	2.06	1.14	2.04	2.02	12.99
U_{15}	1.33	−0.34	3.20	−0.10	−0.59	0.44	0.28	−0.24	−2.83
U_{16}	1.04	−0.71	−1.16	−0.20	−0.97	−1.44	−1.78	0.57	−5.62
U_{17}	0.00	0.12	0.19	−0.71	0.20	0.18	0.18	0.24	−53.45
MD	7.73	8.42	10.29	7.20	10.59	10.56	13.32	14.81	

surement scale. For this case, partial correlations are computed and it was found that they are not significant.

Confirmation Run. After selecting a set of useful variables, a confirmation run is carried out. In the confirmation run, an MS will be constructed for the space generated by the useful variables and MDs of abnormals will also be computed based on these variables. Table 2.5 gives sample values of MDs of the healthy group and abnormal conditions along with their averages. It can be easily seen that the healthy group (with reduced number of variables) has an average MD of 0.975, which is close to 1.0. Interestingly, the average MD of the abnormals is 39.039. This is higher than the average MD (30.39) corresponding to the original variables, which means that insignificant variables would affect the accuracy of the measurement scale.

MTS Method

In MTS, OAs are used to estimate *S/N* ratios of different runs. This method does not provide *S/N* ratios corresponding to the variables directly. In the medical diagnosis case, since there are 17 variables, the $L_{32}(2^{31})$ array is selected. The variables are assigned to the first 17 columns of the array. The presence and absence of the variables are considered as two levels of them. Here, *S/N* ratios are computed based on abnormal MDs only, as they are used for validation of the scale.

The procedure for calculating *S/N* ratios corresponding to a run of an OA is as follows. Let there be t abnormal conditions. Let D_1^2, D_2^2,..., D_t^2, be MDs corresponding to these abnormals. The formula for calculating the *S/N* ratio η_q (for larger-the-better criterion) corresponding the qth run of the OA is

$$\eta_q = -10 \log \left[\frac{1}{t} \sum_{i=1}^{t} \left(\frac{1}{D_i^2} \right) \right] \qquad (2.7)$$

In MTS, *S/N* ratios are calculated with the help of Equation 2.7.

TABLE 2.5   Confirmation Phase of the MTGS Method (Values of MDs)

	1	2	3	4	5	6	7	8	Average
Healthy	0.329	0.441	0.410	0.616	0.420	0.725	0.663	0.546	0.975
Abnormal	10.525	20.465	12.988	8.238	25.089	24.829	32.098	21.159	39.039

In this case also, "dynamic" S/N ratios can be computed if the necessary information is available. Note that Equation 2.7 is given here only to illustrate the process of comparison of these methods. S/N ratios (in dB) along with MDs of five abnormal conditions (actually there are 17 abnormal conditions) for this example are given in Table 2.6.

After computing S/N ratios, average responses are calculated for the variables. A summary of average responses is shown in Table 2.7. From Table 2.7, it is clear that the variables X_2, X_5, X_7,

TABLE 2.6   MDs of Abnormal Data and S/N Ratios (Sample Values)

Run	1	2	3	4	5	S/N Ratio
1	7.73	8.42	10.29	7.21	10.59	11.52
2	8.67	9.49	11.54	8.04	11.92	12.01
3	4.34	6.33	4.06	5.08	4.70	8.23
4	4.62	7.99	4.47	6.52	5.40	8.89
5	2.23	3.37	1.49	2.61	6.26	3.19
6	2.36	4.27	1.83	3.10	7.75	3.25
7	7.50	7.93	12.79	6.87	9.28	9.83
8	8.68	10.01	16.11	7.60	11.29	10.55
9	11.77	5.85	7.61	9.30	5.92	9.51
10	13.91	6.26	8.64	11.65	6.35	9.61
11	8.06	7.49	9.50	4.95	9.37	10.23
12	10.31	9.58	11.90	6.20	11.58	11.14
13	4.49	3.32	4.26	2.67	2.40	6.80
14	5.12	3.98	3.50	3.06	2.94	7.43
15	7.17	1.87	1.90	6.50	5.80	8.70
16	8.46	1.34	2.28	7.77	6.03	8.76
17	4.60	8.94	6.58	2.92	15.27	10.27
18	4.49	9.02	6.66	3.10	15.15	10.29
19	8.39	2.80	9.55	6.62	2.96	6.69
20	7.87	2.81	9.66	7.01	2.93	7.01
21	14.01	7.74	15.28	11.37	12.14	12.36
22	13.80	7.74	15.24	11.50	12.01	12.36
23	7.97	16.32	10.01	6.66	16.56	11.96
24	8.01	16.21	10.31	6.46	15.75	11.96
25	5.30	2.63	4.28	4.15	1.98	6.50
26	5.50	2.74	4.96	4.23	1.96	6.87
27	8.20	4.27	2.91	7.17	5.79	9.76
28	7.50	4.63	2.87	7.27	6.06	9.95
29	12.84	7.28	7.00	10.87	5.83	9.93
30	12.88	7.72	8.01	11.12	6.54	10.52
31	8.97	8.37	9.35	5.48	12.09	10.79
32	8.87	8.39	9.30	5.47	12.06	10.83

TABLE 2.7 Average Responses of All the Variables

	X_1	X_2	X_3	X_4	X_5	X_6	X_7	X_8	X_9	X_{10}	X_{11}	X_{12}	X_{13}	X_{14}	X_{15}	X_{16}	X_{17}
Level 1	8.73	9.40	8.91	9.28	10.74	9.29	9.63	8.90	8.89	9.71	8.76	10.00	9.94	10.26	9.56	9.14	9.24
Level 2	9.88	9.21	9.69	9.33	7.87	9.31	8.98	9.71	9.71	8.89	9.84	8.61	8.66	8.35	9.05	9.46	9.37
Gain	-1.15	0.19	-0.78	-0.05	2.88	-0.02	0.65	-0.80	-0.82	0.82	-1.08	1.39	1.28	1.91	0.51	-0.32	-0.13

X_{10}, X_{12}, X_{13}, X_{14}, and X_{15} are useful because they have positive gains. Both methods gave almost the same useful variable combinations: X_2-X_5-X_6-X_7-X_{10}-X_{12}-X_{13}-X_{14}-X_{15} by MTGS and X_2-X_5-X_7-X_{10}-X_{12}-X_{13}-X_{14}-X_{15} by MTS.

Confirmation Run. A confirmation run was conducted with the set of useful variables. Here, MS will be generated with useful variables. The MDs corresponding to abnormals are obtained with useful variables. Sample results of the confirmation run are given in Table 2.8. The average MD of the healthy group converges to 1.0 in this case also. However, the average MD of abnormals is 45.553. This value is more than that obtained in the confirmation phase of MTGS, which is 39.039.

In MTGS the useful variables are identified based on the Gram-Schmidt vectors. Since these vectors are orthogonal, we can obtain the effects of the variables independently. Therefore, MTGS saves some computational time. Note that if the effects of partial correlations are significant and a definite order for the variables cannot be defined, then OAs have to be used in MTGS. In that case, the results of the two methods would be the same.

2.6 CONCLUSIONS

- MTS and MTGS are different methods that use Taguchi Methods and Mahalanobis distance to diagnose multivariate systems.
- Mahalanobis space defined in these methods is extremely useful, since we can generalize the definition of MS to any number of variables.

TABLE 2.8 Confirmation Run of the MTS Method (Values of MDs)

	1	2	3	4	5	6	7	8	Average
Healthy	0.329	0.651	0.375	0.463	0.482	0.717	0.555	0.491	0.995
Abnormal	13.939	15.457	19.051	11.523	19.710	18.665	24.085	29.149	45.553

- To define the Mahalanobis space we do not need to assume any distribution for the input variables. This statement is proved in this chapter.
- MTGS is simpler than MTS if partial correlation effects are not significant and there exists a definite order of the variables.

3

ADVANTAGES AND LIMITATIONS OF MTS AND MTGS

In this chapter the relative advantages and limitations of MTS and MTGS methods are discussed in terms of the direction of abnormals, multicollinearity problems, and partial correlations.

3.1 DIRECTION OF ABNORMALITIES

One of the main reasons for using Mahalanobis distance in multivariate diagnosis is its ability to distinguish abnormalities from "healthy" or "normal" group. Sometimes the abnormal conditions arise out of extremely good situations. Therefore, it is important to identify the direction of abnormalities to enable the distinction between "good" and "bad" abnormalities. This makes the diagnosis process more effective. If MD is calculated using the inverse of the correlation matrix (MTS method), such distinctions cannot be made. However, this can be accomplished if we use the Gram–Schmidt orthogonalization process to calculate MD (MTGS method). This chapter outlines a procedure to identify the direction of abnormals with suitable equations for different cases. The use of this procedure is demonstrated for a graduate student admission system. Since the Gram–Schmidt orthogonalization process is used to identify the directions of abnormals, it is worthwhile to review this process again.

3.1.1 The Gram–Schmidt Process

Gram–Schmidt process can be simply stated: Given linearly independent vectors $Z_1, Z_2, ..., Z_k$, there exist mutually perpendicular vectors $U_1, U_2, ..., U_k$ with the same linear span. This process is described in the Figure 3.1, which is the same as Figure 2.2. The construction of orthogonal vectors and the computation of MD with these vectors are explained in the Section 2.2.2.

3.1.2 Identification of the Direction of Abnormals

The GSP vectors obtained in the Section 2.2.2 can be written as

$$U_1 = Z_1$$

$$U_2 = Z_2 - c_{2,1}U_1$$

$$U_3 = Z_3 - c_{3,1}U_1 - c_{3,2}U_2$$

$$\vdots$$

$$U_k = Z_k - c_{k,1}U_1 - c_{k,2}U_2 - \cdots - c_{k,k-1}U_{k-1}$$

where $c_{2,1}, c_{3,1}, ..., c_{k,k-1}$ are Gram–Schmidt vector coefficients.

First, a discussion on the direction of abnormals is presented in detail for a two-variable case. The same logic is then extended for higher number of variables. In the case of two variables, the distribution of the points forms an elliptical shape. Since MS is constructed based on this distribution, for simplicity, we can represent this distribution as the MS. The elliptical shape remains unchanged after orthogonal transformation; therefore, the distribution corresponding to GSP vectors will also have an elliptical shape.

We know that the mean of the GSP vectors is located at the zero point. For the two-variable case, the mean of U_1 and U_2 is located at (0, 0). All conditions (normal and abnormal) are above

Figure 3.1 Gram–Schmidt process.

or below this point with the exception of the points that match with the zero point (mean). It is therefore clear that the abnormals are above or below the zero point. Based on the position of abnormals and the value of MD, we can distinguish between "good" and "bad" abnormals. In the case of two variables, depending on the types of characteristics, we have the following four cases:

1. Both U_1 and U_2 are larger-the-better type
2. U_1 is smaller-the-better type and U_2 is larger-the-better type
3. U_1 is larger-the-better type and U_2 is smaller-the-better type
4. Both U_1 and U_2 are smaller-the-better type

For a two-variable case the vectors U_1 and U_2 can be written as

$$U_1 = Z_1$$
$$U_2 = Z_2 - c_{2,1}U_1$$

or

$$(u_{11}, u_{12}, ..., u_{1n}) = (z_{11}, z_{12}, ..., z_{1n})$$
$$(u_{21}, u_{22}, ..., u_{2n}) = (z_{21} - c_{2,1}u_{11}, z_{22} - c_{2,1}u_{12}, ..., z_{2n} - c_{2,1}u_{1n})$$

For all four cases, the rules for identifying the direction of abnormals are described below.

Case 1: Both U_1 and U_2 Are Larger-the-Better Type

An example where the characteristics U_1 and U_2 are of larger-the-better type is a student admission system. In this system U_1 could be "GPA" (grade-point average) and U_2 could be "TOEFL exam score." In this case, a condition corresponding to very low U_1 and U_2 is an abnormal, and a condition corresponding to very high U_1 and U_2 is also an abnormal.

For the jth condition to be a good abnormal, the elements of U_1 and U_2 should be above zero (positive) and the corresponding MD should be larger than threshold T. These conditions can be mathematically represented as

1. $u_{1j} > 0$ or $z_{1j} > 0$; $u_{2j} > 0$ or $z_{2j} - c_{2,1} u_{1j} > 0$ or $z_{2j} > c_{2,1} u_{1j}$
2. $MD_j > T$

From Equation 2.3, we can write

$$\frac{1}{2} \left(\frac{u_{1j}^2}{s_1^2} + \frac{u_{2j}^2}{s_2^2} \right) > T \quad \text{or} \quad \frac{u_{1j}^2}{s_1^2} + \frac{u_{2j}^2}{s_2^2} > 2T$$

The pictorial representation of these rules is given in Figure 3.2. For this case the decision rule can be stated as follows: If elements of U_1 and U_2 corresponding to an abnormal condition are positive and MD is higher than the threshold T, then the abnormal condition can be classified as good abnormality; otherwise, it is a bad abnormality.

Case 2: U_1 Is Smaller-the-Better Type and U_2 Is Larger-the-Better Type

An example of this situation is the banking system to grant loans. Here U_1 could be the "number of people in a household" and U_2 could be the "income level of a family." In this case, a condition corresponding to very high U_1 and very low U_2 is an abnormal and a condition corresponding to very low U_1 and very high U_2 is also an abnormal.

For the jth condition to be a good abnormal, the element of U_1 should be below zero (negative) and that of U_2 should be above zero (positive) and the corresponding MD should be larger than threshold T. These conditions can be mathematically represented as

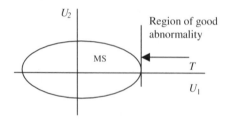

Figure 3.2 Both U_1 and U_2 are larger-the-better type.

1. $u_{1j} < 0$ or $z_{1j} < 0$; $u_{2j} > 0$ or $z_{2j} - c_{2,1} u_{1j} > 0$ or $z_{2j} > c_{2,1} u_{1j}$
2. $MD_j > T$

From Equation 2.3, we can write

$$\frac{1}{2}\left(\frac{u_{1j}^2}{s_1^2} + \frac{u_{2j}^2}{s_2^2}\right) > T \quad \text{or} \quad \frac{u_{1j}^2}{s_1^2} + \frac{u_{2j}^2}{s_2^2} > 2T$$

The pictorial representation of these rules is given in Figure 3.3. For this case the decision rule can be stated as follows: If the element of U_1 is negative and that of U_2 is positive and MD is higher than the threshold T, then the abnormal condition can be classified as good abnormality; otherwise, it is a bad abnormality.

Case 3: U_1 Is Larger-the-Better Type and U_2 Is Smaller-the-Better Type

An example of this situation is an inspection system, where the characteristic "tensile strength" (U_1) is larger-the-better type and the characteristic "number of occurrences of a particular defect" (U_2) is smaller-the-better type. In this case, a condition corresponding to very low U_1 and very high U_2 is an abnormal and a condition corresponding to very high U_1 and very low U_2 is also an abnormal. For jth condition to be a good abnormal, the element of U_1 should be above zero (positive) and that of U_2 should be below zero (negative) and the corresponding MD should be larger than threshold T. These conditions can be mathematically represented as

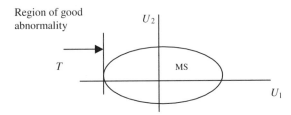

Figure 3.3 U_1 is smaller-the-better type and U_2 is larger-the-better type.

1. $u_{1j} > 0$ or $z_{1j} > 0$; $u_{2j} < 0$ or $z_{2j} - c_{2,1} u_{1j} < 0$ or $z_{2j} > c_{2,1} u_{1j}$
2. $MD_j > T$

From Equation 2.3, we can write

$$\frac{1}{2}\left(\frac{u_{1j}^2}{s_1^2} + \frac{u_{2j}^2}{s_2^2}\right) > T \quad \text{or} \quad \frac{u_{1j}^2}{s_1^2} + \frac{u_{2j}^2}{s_2^2} > 2T$$

The pictorial representation of these rules is given in Figure 3.4. For this case the decision rule can be stated as follows: If the element of U_1 is positive and that of U_2 is negative and MD is higher than the threshold, T then the corresponding abnormal condition can be classified as good abnormality; otherwise, it is a bad abnormality.

Case 4: Both U_1 and U_2 Are Smaller-the-Better Type

An example of a situation where both the characteristics are of the smaller-the-better type is a printed circuit board inspection system. In this example, the characteristics "number of occurrences of a particular defect" (U_1) and "line width reduction after etching process" (U_2) are of smaller-the-better type. In this case a condition corresponding to very high U_1 and U_2 is an abnormal and a condition corresponding to very low U_1 and U_2 is also an abnormal.

For the jth condition to be a good-abnormal, the elements of U_1 and U_2 should be below zero (negative) and the corresponding MD should be larger than threshold T. These conditions can be mathematically represented as

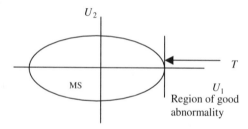

Figure 3.4 U_1 is larger-the-better type and U_2 is smaller-the-better type.

1. $u_{1j} < 0$ or $z_{1j} < 0$; $u_{2j} < 0$ or $z_{2j} - c_{2,1}\, u_{1j} < 0$
   or $z_{2j} < c_{2,1}\, u_{1j}$
2. $\mathrm{MD}_j > T$

From Equation 2.3, we can write

$$\frac{1}{2}\left(\frac{u_{1j}^{\,2}}{s_1^{\,2}} + \frac{u_{2j}^{\,2}}{s_2^{\,2}}\right) > T \quad \text{or} \quad \frac{u_{1j}^{\,2}}{s_1^{\,2}} + \frac{u_{2j}^{\,2}}{s_2^{\,2}} > 2T$$

The pictorial representation of these rules is given in Figure 3.5. For this case the decision rule can be stated as follows: If the elements of U_1 and U_2 corresponding to an abnormal condition are negative and MD is higher than the threshold T, then the abnormal condition can be classified as good abnormality; otherwise, it is a bad abnormality.

3.1.3   Decision Rule for Higher Dimensions

If there are k variables and the sample size is n, the conditions for jth abnormal to be good are

$$u_{1j} > 0, \quad \text{if } U_1 \text{ is larger-the-better type}$$

$$(<0, \text{ if } U_1 \text{ is smaller-the-better type})$$

$$u_{2j} > 0, \quad \text{if } U_2 \text{ is larger-the-better type}$$

$$(<0, \text{ if } U_2 \text{ is smaller-the-better type})$$

$$\vdots$$

$$u_{kj} > 0, \quad \text{if } U_k \text{ is larger-the-better type}$$

$$(<0, \text{ if } U_k \text{ is smaller-the-better type})$$

and

$$\frac{1}{k}\left(\frac{u_{1j}^{\,2}}{s_1^{\,2}} + \frac{u_{2j}^{\,2}}{s_2^{\,2}} + \cdots + \frac{u_{kj}^{\,2}}{s_k^{\,2}}\right) > T,$$

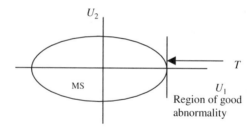

Figure 3.5   Both U_1 and U_2 are smaller-the-better type.

or

$$\frac{u_{1j}^2}{s_1^2} + \frac{u_{2j}^2}{s_2^2} + \cdots + \frac{u_{kj}^2}{s_k^2} > kT$$

otherwise, the abnormal condition is bad.

Though this procedure provides guidelines to identify the direction of abnormals, it is left to the decision maker to decide the type of abnormal conditions. For example, in a student admission system, where the variables are mostly of larger-the-better type, a student with very high scores on most of the variables can be considered as a good abnormal by decision maker. The above method only provides a means to identify the direction of abnormals. It is left to decision maker to make a decision on the type of abnormals, depending on the need.

This procedure is applicable to the cases where we have knowledge about the type of variables (like larger-the-better type or smaller-the-better type). Development of a methodology to identify the direction of abnormals for the cases where the type of variables cannot be specified (for example, cases with categorical data) can be a topic for future research.

3.2   EXAMPLE OF A GRADUATE ADMISSION SYSTEM

The applicability of the above procedure can be demonstrated with a university student admission system. Let us suppose that a student is admitted into the university based on the following three variables.

- *Grade-point average (GPA)*: This should be as high as possible on a 4.0 scale.
- *GMAT score:* This should be as high as possible (maximum score is 1600).
- *SAT-Math score:* This should be as high as possible (maximum score is 800).

The data corresponding to the healthy group along with GSP vectors and MDs are shown in Table 3.1. Since these variables are larger-the-better type, this example will fall into the first category (case 1) of Section 3.1.2. There are eight abnormal conditions $A_1,...,A_8$. The abnormal data along with GSP vectors and MDs are given in Table 3.2.

From Table 3.2, it is clear that all of the abnormalities have high MDs. Based on MDs alone we cannot distinguish between good and bad abnormalities. To make such distinctions we need to look at GSP vectors. Assuming that the threshold T is set at 3.0, A_7 and A_8 can be classified as good abnormalities because GSP vectors corresponding to these abnormalities have positive signs and their MDs are higher than threshold. The remaining

TABLE 3.1 Healthy Group Data

	Original Variables			Gram–Schmidt Vectors			
S. No.	X_1 (GPA)	X_2 (SAT)	X_3 (SAT-Math)	U_1	U_2	U_3	MD
1	3.0	1010	670	−0.465	−0.158	0.951	0.537
2	2.9	990	428	−0.714	−0.081	−1.775	1.703
3	3.2	1035	712	0.033	−0.409	1.322	1.028
4	2.8	980	546	−0.963	0.060	−0.355	0.374
5	3.9	1310	677	1.776	−0.062	−0.103	1.061
6	3.2	990	650	0.033	−0.703	0.749	0.807
7	2.8	965	646	−0.963	−0.038	0.854	0.664
8	3.7	1380	715	1.278	0.810	0.144	1.264
9	3.4	1300	645	0.531	0.908	−0.355	1.048
10	3.2	1205	645	0.033	0.702	−0.010	0.534
11	2.6	895	490	−1.461	−0.081	−0.693	0.951
12	3.1	950	555	−0.216	−0.757	−0.205	0.656
13	3.6	1110	520	1.029	−0.747	−1.220	1.679
14	2.7	1045	625	−1.212	0.692	0.367	1.074
15	3.7	1235	690	1.278	−0.138	0.327	0.617

TABLE 3.2   Abnormal Data

Abnormal Condition	Original Variables			Gram–Schmidt Vectors			MD
	X_1 (GPA)	X_2 (GMAT)	X_3 (SAT-Math)	U_1	U_2	U_3	
A_1	2.40	1210	540	−1.959	2.392	−1.105	8.064
A_2	1.80	765	280	−3.453	0.727	−2.566	7.740
A_3	0.90	540	280	−5.695	1.122	−1.676	13.532
A_4	3.60	990	230	1.029	−1.532	−4.194	11.417
A_5	2.10	930	480	−2.706	1.184	−0.836	4.297
A_6	2.6	1140	530	−1.461	1.520	−1.028	3.725
A_7	4	1600	800	2.026	1.626	0.361	4.291
A_8	3.9	1580	780	1.776	1.702	0.212	4.210

abnormalities (A_1–A_6) can be classified as bad abnormalities because not all elements of GSP vectors have positive signs even though their MDs are higher than threshold.

3.3   MULTICOLLINEARITY

The problem of strong relationships between variables is commonly referred to as multicollinearity. In such situations, the correlation matrix becomes almost singular* and the inverse matrix is not accurate. When there are strong relationships between the variables, MTGS gives fairly good MDs as compared to MTS. This is illustrated with a data set, which is provided by American Supplier Institute (ASI). The data set corresponds to the healthy group. This group has observations on 13 variables with a sample size of 44. The complete data are shown in Appendix 1. MTS and MTGS methods are used to compute MDs for these data. From the correlation matrix (Table 3.3), it is clear that the variables are highly correlated indicating existence of multicollinearity.

Table 3.4 gives the inverse of the correlation matrix. The determinant of this matrix is calculated and found to be −4.66 × 10^{-15}, indicating that the matrix is almost singular. By using this

---

* A matrix is said to be singular if its determinant is zero. A determinant is a number associated with a square matrix.

TABLE 3.3 Correlation Matrix

	X_1	X_2	X_3	X_4	X_5	X_6	X_7	X_8	X_9	X_{10}	X_{11}	X_{12}	X_{13}
X_1	1.000	0.833	1.000	0.801	1.000	0.911	0.789	0.931	0.826	0.924	0.880	0.962	0.795
X_2	0.833	1.000	0.816	0.823	0.811	0.930	0.950	0.862	0.928	0.864	0.938	0.919	0.949
X_3	1.000	0.816	1.000	0.756	0.988	0.881	0.771	0.883	0.808	0.876	0.850	0.914	0.777
X_4	0.801	0.823	0.756	1.000	0.795	0.722	0.888	0.908	0.847	0.949	0.792	0.883	0.905
X_5	1.000	0.811	0.988	0.795	1.000	0.859	0.784	0.901	0.818	0.900	0.840	0.924	0.790
X_6	0.911	0.930	0.881	0.722	0.859	1.000	0.850	0.831	0.829	0.813	0.953	0.905	0.845
X_7	0.789	0.950	0.771	0.888	0.784	0.850	1.000	0.877	0.953	0.899	0.887	0.899	0.977
X_8	0.931	0.862	0.883	0.908	0.901	0.831	0.877	1.000	0.865	0.945	0.861	0.935	0.891
X_9	0.826	0.928	0.808	0.847	0.818	0.829	0.953	0.865	1.000	0.881	0.847	0.890	0.954
X_{10}	0.924	0.864	0.876	0.949	0.900	0.813	0.899	0.945	0.881	1.000	0.854	0.941	0.913
X_{11}	0.880	0.938	0.850	0.792	0.840	0.953	0.887	0.861	0.847	0.854	1.000	0.920	0.888
X_{12}	0.962	0.919	0.914	0.883	0.924	0.905	0.899	0.935	0.890	0.941	0.920	1.000	0.910
X_{13}	0.795	0.949	0.777	0.905	0.790	0.845	0.977	0.891	0.954	0.913	0.888	0.910	1.000

TABLE 3.4 Inverse Matrix

	X_1	X_2	X_3	X_4	X_5	X_6	X_7	X_8	X_9	X_{10}	X_{11}	X_{12}	X_{13}
X_1	-8.359	-0.387	6.304	0.723	2.785	0.275	0.095	0.033	-0.350	-0.358	-0.022	0.601	-0.965
X_2	-0.387	31.175	0.828	0.407	3.583	-11.662	-6.593	0.326	-5.435	3.258	-3.226	-4.800	-7.664
X_3	6.304	0.828	60.005	7.352	-58.951	-10.767	7.373	-1.918	-4.744	-6.073	-1.377	1.585	0.955
X_4	0.723	0.407	7.352	17.836	-4.707	3.557	-1.952	-3.668	2.719	-12.540	-0.336	-2.618	-5.400
X_5	2.785	3.583	-58.951	-4.707	65.584	2.117	-2.374	-2.499	-3.392	-4.323	2.091	-8.674	8.604
X_6	0.275	-11.662	-10.767	3.557	2.117	23.310	-2.691	0.580	4.736	3.115	-9.520	-2.594	0.184
X_7	0.095	-6.593	7.373	-1.952	-2.374	-2.691	30.611	-0.281	-8.899	-4.653	-1.578	2.283	-11.227
X_8	0.033	0.326	-1.918	-3.668	-2.499	0.580	-0.281	12.988	0.837	-1.633	-0.674	-2.073	-2.158
X_9	-0.350	-5.435	-4.744	2.719	-3.392	4.736	-8.899	0.837	19.315	2.680	3.875	0.509	-11.483
X_{10}	-0.358	3.258	-6.073	-12.540	-4.323	3.115	-4.653	-1.633	2.680	29.545	-0.618	-3.683	-5.569
X_{11}	-0.022	-3.226	-1.377	-0.336	2.091	-9.520	-1.578	-0.674	3.875	-0.618	16.825	-2.548	-2.765
X_{12}	0.601	-4.800	1.585	-2.618	-8.674	-2.594	2.283	-2.073	0.509	-3.683	-2.548	23.211	-2.099
X_{13}	-0.965	-7.664	0.955	-5.400	8.604	0.184	-11.227	-2.158	-11.483	-5.569	-2.765	-2.099	39.515

inverse matrix, MDs for the healthy group are computed (MTS method). For the same data set MDs are computed using the Gram–Schmidt method (MTGS method). A graphical representation of MDs obtained by these two methods is shown in Figure 3.6.

From Figure 3.6, it is clear that by using MTGS we can obtain fairly good values of MDs for the healthy group. The average value of MD of the healthy group in the case of MTGS is 0.977, which is the desired value, whereas the corresponding average using MTS is 0.56.

Note. There is another way of handling multicollinearity, without using MTGS. In this method, we propose the use of the adjoint of the correlation matrix to calculate the distances. The details of this method are discussed in Chapter 9.

3.4 A DISCUSSION OF PARTIAL CORRELATIONS

When the MTGS method is used for diagnosis, we can estimate the effects of the variables independently without using OAs only if the effects of partial correlations are not significant. Though GSP variables $(U_1,U_2,...,U_k)$ are orthogonal, there might exist a relation between the GSP variables and original variables $(Z_1,Z_2,...,Z_k)$. Therefore, it is necessary to estimate the association between $U_1,U_2,...,U_k$ and $Z_1,Z_2,...,Z_k$. Such effects can be measured using partial correlation coefficients. Morrison (1967) provided a detailed discussion on partial correlations.

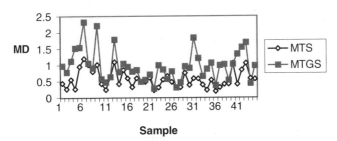

Figure 3.6 MDs from the MTS and MTGS methods for ASI data.

Computation of Partial Correlation Coefficients

Consider the following errors:

$$U_1 - \mu_{U_1} - S_{U_1 Z} S^{-1}_{ZZ}(Z - \mu_Z)$$

$$U_2 - \mu_{U_2} - S_{U_2 Z} S^{-1}_{ZZ}(Z - \mu_Z)$$

$$\vdots$$

$$U_k - \mu_{U_k} - S_{U_k Z} S^{-1}_{ZZ}(Z - \mu_Z)$$

These errors are obtained from the Gram–Schmidt orthogonalization process used to compute the vectors U_1, U_2,...,U_k. Their correlations are determined from the error covariance matrix $S_{UU.Z} = S_{UU} - S_{UZ} S_{ZZ}^{-1} S_{ZU}$, which measures the association between GSP vectors after eliminating the effects of $Z_1, Z_2,...,Z_k$.

In the above equation, the S_{UU} is a covariance matrix with the covariance between GSP vectors as its elements, S_{UZ} is a covariance matrix with covariance between GSP vectors and original vectors as its elements, and so on.

The partial correlation between U_1 and U_2, eliminating the effects of $Z_1, Z_2,...,Z_k$, is given by

$$r_{U_1 U_2.Z} = \frac{S_{U_1 U_2.Z}}{\sqrt{S_{U_1 U_1.Z}} \sqrt{S_{U_2 U_2.Z}}} \tag{3.1}$$

where $S_{U_1 U_2.Z}$ represents partial covariance between U_1 and U_2, eliminating the effects of $Z_1, Z_2,...,Z_k$ and the terms under the square root signs represent variances of U_1 and U_2. Similarly, other partial correlation coefficients can be obtained. $S_{U_i U_j.Z}$ is the (i, j) element of the matrix $S_{UU} - S_{UZ} S_{ZZ}^{-1} S_{ZU}$. The matrix containing the partial correlation coefficients can be written as

$$\text{Partial correlation matrix} = D^{-1/2}(S_{UU} - S_{UZ} S_{ZZ}^{-1} S_{ZU}) D^{-1/2}$$

$$\tag{3.2}$$

where the diagonal matrix $D = \text{diag.}(S_{UU} - S_{UZ} S_{ZZ}^{-1} S_{ZU})$, and $D^{-1/2}$ is a matrix consisting of square root of reciprocals of elements of D.

3.5 CONCLUSIONS

- It is important to identify the direction of abnormals in multivariate systems, such as a student admission system or medical diagnosis. The decision maker can treat the good-abnormality group separately. In the case of a student admission system the good-abnormality group can be awarded scholarships/fellowships and in the case of medical diagnosis system the doctor can reduce the medical expenses of the patients in this group by reducing the frequency of checkups.

- It has been proved that with the help of the Gram–Schmidt method (MTGS method), we can determine the direction of the abnormalities, which will enable a decision maker to take appropriate actions. This is not possible by the inverse correlation matrix method (MTS method). The procedure to identify the direction of abnormals is limited to the cases where we have knowledge about the type of variables. Development of a methodology for the cases where type of var-

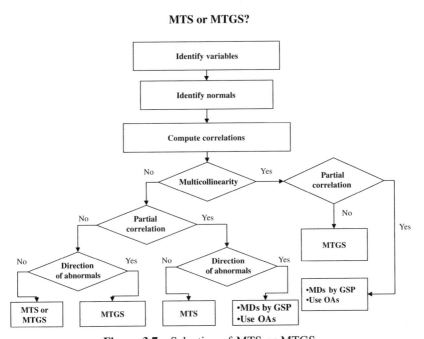

Figure 3.7   Selection of MTS or MTGS.

iables cannot be specified (for example, cases with categorical data) is a good area for future research.

- The MTGS method gives fairly good results in cases where multicollinearity problems are present.
- If the effects of the partial correlations are not strong, then there is no need to use OAs to compute S/N ratios and hence the useful variables in the MTGS method. However, use of OAs is recommended, since one need not be concerned about the effects of the partial correlations etc.
- The question of choosing MTS/MTGS depends on factors such as the direction of the abnormals, multicollinearity, and partial correlations. Figure 3.7 helps in choosing one of the two methods (MTS or MTGS) based on these factors.

4

ROLE OF ORTHOGONAL ARRAYS AND SIGNAL-TO-NOISE RATIOS IN MULTIVARIATE DIAGNOSIS

In multivariate systems, the dimensionality reduction is still a challenge. A higher number of dimensions restrict the applicability of several multivariate techniques to obtain meaningful results. In this chapter the applicability of orthogonal arrays (OAs) and signal-to-noise (S/N) ratios for dimensionality reduction is demonstrated. In robust engineering, OAs are used to estimate the effects of several factors and the effect of interactions by minimizing the number of experiments. A good discussion on orthogonal arrays is given in Taguchi (1987). For each experiment S/N ratios are computed. The S/N ratio is a measure of the functionality of the system, which exploits interaction between control factors and noise factors. A gain in S/N ratio indicates a reduction in the variability, which will result in cost savings if we employ the quality loss function approach. Phadke (1989) and Park (1996) give a good description of the different types of S/N ratios. As in robust engineering, the S/N ratio plays an important role in MTS/MTGS. The measure can be used for the evaluation of a multi-dimensional system and for minimizing the number of variables of the system.

4.1 ROLE OF ORTHOGONAL ARRAYS

In robust engineering, the main role of OAs is to permit engineers to evaluate a product design with respect to robustness against noise and cost involved. The OA is an inspection device to prevent a "poor design" from going "downstream."

Usually, these arrays are denoted as L_a (b^c), where

a = the number of experimental runs
b = the number of levels of each factor
c = the number of columns in the array
L = denotes Latin square design

Arrays can have factors with many levels, although two- and three-level factors are most commonly encountered.

In MTS/MTGS, OAs are used to select the variables of importance by minimizing the different combinations of the original set of variables. The variables are assigned to the different columns of the array. The presence and absence of the variables are considered as the levels. Since the variables have only two levels, two-level arrays are used in MTS/MTGS applications. The importance of the variables is judged based on their ability to measure the degree of abnormality on the measurement scale. For each run of an OA, MDs corresponding to the known abnormal conditions or the conditions outside MS are computed. We need not consider the MDs corresponding to the MS (healthy group), because we know that this group is healthy and the scale is constructed based on this group.

Let us consider an example. Let there be five variables X_1, X_2, X_3, X_4, and X_5 allocated to the first five columns of an L_8 (2^7) array, as shown in Table 4.1. In this table, 1 indicates the level corresponding to the presence of a variable and 2 indicates the level corresponding to the absence of a variable. Consider the first three runs of Table 4.1.

1st Run: 1-1-1-1-1

In this run, MS is constructed with the variable combination X_1-X_2-X_3-X_4-X_5. The correlation matrix is of the order 5 × 5. The MDs of the abnormals are estimated with this matrix.

TABLE 4.1   Variable Allocation in L_8 (2^7) Array

L_8 (2^7) Array Column Run/variable	1 X_1	2 X_2	3 X_3	4 X_4	5 X_5	6	7
1	1	1	1	1	1	1	1
2	1	1	1	2	2	2	2
3	1	2	2	1	1	2	2
4	1	2	2	2	2	1	1
5	2	1	2	1	2	1	2
6	2	1	2	2	1	2	1
7	2	2	1	1	2	2	1
8	2	2	1	2	1	1	2

2nd Run: 1-1-1-2-2

In this run MS is constructed with the variable combination X_1-X_2-X_3. The correlation matrix is of the order 3 × 3. The MDs of abnormals are estimated with this matrix.

3rd Run: 1-2-2-1-1

In this run MS is constructed with the variable combination X_1-X_4-X_5. The correlation matrix is of the order 3 × 3. The MDs of abnormals are estimated with this matrix.

Based on MDs of the conditions, outside MS, corresponding to the runs of an orthogonal array, we can evaluate *S/N* ratios. Thus, OAs are required for testing a smaller number of different combinations of the variables to find out the variables of importance based on their ability to measure the conditions outside MS. However, OAs are not necessary in MTGS method if the effects of partial correlations are not significant, since we can get the estimates directly from the orthogonal variables.

4.2   ROLE OF *S/N* RATIOS

The signal to noise ratio tries to capture the magnitude of true information (i.e., signals) after making some adjustment for uncontrollable variation (i.e., noise). In the case of multivariate diagnosis, *S/N* ratio is defined as the measure of accuracy of the measurement scale for predicting the abnormal conditions.* S/N ratio is expressed in decibels (dB). A higher value of the S/N ratio means lower error of prediction. S/N ratios are calculated using abnormal conditions only, since it is necessary to test the accuracy of the scale based on the conditions outside MS.

In multidimensional applications, it is important to identify a useful set of variables that is sufficient to detect the abnormals. It is also important to assess the performance of the given system and the degree of improvement in the performance. In MTS/ MTGS methods, S/N ratios are used to accomplish these objectives.

After obtaining the MDs for the known abnormal conditions corresponding to the various combinations of an OA, S/N ratios are computed for all these combinations to determine the useful set of variables. S/N ratios are important to improve the accuracy of the measurement scale and reduce the cost of diagnosis. The useful set of variables is obtained by evaluating the "gain" in S/N ratios. The gain in S/N ratio is the difference between the average S/N ratio when the variable is used in OA and the average S/N ratio when the variable is not used in OA. If the gain is positive, then the variable is useful.

By using S/N ratios, performance of the entire system can be measured. S/N ratios corresponding to all of the variables (original combination, which is usually one of the combinations of the OA) and the useful set of variables (optimal combination, which may not be one of the combinations of the OA) are computed based on MDs of abnormal conditions. The gain in S/N ratio indicates the degree of improvement in the diagnosis process with respect to the original system.

The S/N ratios can also be used to assess the performance of the system under different conditions. Using S/N ratios, a useful set of variables can be obtained for different conditions. This helps in finding the important variables for different diseases or the important variables required for selling a product in different regions.

From this discussion, the advantages of S/N ratios can be summarized:

---

*These conditions need not be abnormal conditions. In fact, any condition outside MS can be used. This holds good for all relevant discussions in the book.

1. They provide a simple way to identify a set of useful variables.
2. They can measure the functionality of a system.
3. They can assess the performance of a condition with different abnormals and find a useful set of variables for this condition.

Types of *S/N* Ratios

In MTS/MTGS methods, typically two types of *S/N* ratios are used: (1) the *larger-the-better* type, and (2) the *dynamic* type. If known abonormals are used as conditions outside MS and the true levels of these abnormals are not known, larger-the-better type *S/N* ratios are used because, for abnormals the distances should be large. When the levels of abnormals are known, dynamic *S/N* ratios are used.

In some cases, it is quite difficult to obtain the levels of severity of the abnormals based on the knowledge of the person conducting the diagnosis. In such situations, if we know the abnormals with different degrees of severity (different categories), dynamic *S/N* ratios can be computed by taking the average of the square root of MDs (working average) in each category as known levels of abnormals. Though the larger-the-better type *S/N* ratio is used in some cases, it is always better to use dynamic type of *S/N* ratios (Taguchi 1987).

Equations for S/N Ratios (MTS Method)

In this section equations for both types of *S/N* ratios are developed.

The Larger-the-Better Type. If the true levels of severity are not known, the larger-the-better type *S/N* ratios are used. Examples for this type are medical diagnosis and manufacturing inspection. The objective of using *S/N* ratios is to select the variables that are sufficient to identify the abnormal conditions. If abnormal MDs are used as the responses of OA, MDs corresponding to them should have larger values. In such cases, it is recommended to use larger-the-better type *S/N* ratios.

The procedure for calculating S/N ratios corresponding to a run of an OA is as follows. Let there be t abnormal conditions. Let D_1^2, D_2^2,..., D_t^2, be MDs corresponding to these abnormal situations. The S/N ratio (for larger-the-better criterion) corresponding to the qth run of OA is given by

$$S/N \text{ ratio} = \eta_q = -10 \log \left[\frac{1}{t} \sum_{i=1}^{t} \left(\frac{1}{D_i^2} \right) \right] \qquad (4.1)$$

The Dynamic Type. Dynamic systems are those in which we want the system's response to follow the levels of the signal factor (see Figure 1.2) in a prescribed manner. The S/N ratios computed for these systems are referred to as the dynamic S/N ratios. This section describes the procedures of computing the dynamic S/N ratios. The dynamic S/N ratios are calculated in the following two types of situations:

1. If the true levels of the severity of all the abnormals are known
2. If the true levels of the severity are not known and working averages are used

CASE 1: THE TRUE LEVELS OF THE SEVERITY OF ALL THE ABNORMALS ARE KNOWN.   Examples for this type are weather forecasting and rainfall prediction. In the case where rainfalls are known, we are fitting a relation between y = the square root of Mahalanobis distances and the Ms (= rainfall figures) and using S/N ratio as the measure of variability. For the discussions on dynamic-type S/N ratios, let us refer to Figure 1.2. If $M_1, M_2,...,M_t$ represent the true levels of severity (input signals) corresponding to t abnormals, the relationship between the input signal (M_is) and the output response (ys) is given by

$$y_i = \beta M_i \qquad (4.2)$$

where $y_i = \sqrt{MD_i}$, $i = 1,...,t$, and β is the slope.

The computational procedure for S/N ratios is described as follows. For each run of the OA construct an ANOVA table:

Source	Degrees of Freedom	Sum of Squares
β	1	S_β
e	$t - 1$	S_e
Total	T	S_T

where

S_T = total sum of squares = $\Sigma_{i=1}^t y_i^2$
r = sum of squares due to input signal = $\Sigma_{i=1}^t M_i^2$
S_β = sum of squares due to slope = $(1/r)(\Sigma_{i=1}^t M_i y_i)^2$
S_e = error sum of squares = $S_T - S_\beta$
V_e = error variance = $S_e/(t - 1)$

The *S/N* ratio of the corresponding qth run of OA is given by

$$S/N \text{ ratio} = \eta_q = 10 \log \left(\frac{1}{r} \frac{S_\beta - V_e}{V_e} \right) \qquad (4.3)$$

The estimate of the slope β for the qth run of the OA is given by

$$\beta_q^2 = \frac{1}{r} (S_\beta - V_e) \qquad (4.4)$$

CASE 2: DYNAMIC *S/N* RATIOS WITH WORKING AVERAGES.   In some cases, if the true levels of abnormals are not known, it is better to compute dynamic *S/N* ratios using working averages. Dynamic type *S/N* ratios are preferred to the larger-the-better type because dynamic type *S/N* ratios would give more accurate predictions (Taguchi 1987). Let there be k classes of abnormals with different degrees of severity. The relationship between the input signal (M_is) and the output response (**y**s) is given by

$$\mathbf{y}_i = \beta \mathbf{M}_i \qquad (4.5)$$

where $y_i = \sqrt{MD}$, $i = 1,...,k$, and β is the slope.
    M_i = working mean of the ith class ($i = 1,...,k$)

In Equation 4.5, the slope β is unity, because the input M_i is the average of square root of MDs in ith class. If each class has $\underline{m}$ observations (MDs), then $M_i = (1/m)(\Sigma_{j=1}^{m} y_{ij})$, where $y_{ij} = \sqrt{MD}$, $i = 1,...,k$, and $j = 1,...,m$.

The computational procedure for S/N ratios is

$$S_T = \text{total sum of squares} = \sum_{i=1}^{k} \sum_{j=1}^{m} y_{ij}^2$$

$$r = \text{sum of squares due to input signal} = m\left(\sum_{i=1}^{k} M_i^2\right)$$

$$S_\beta = \text{sum of squares due to slope} = \left(\frac{1}{r}\right)\left(\sum_{i=1}^{k} M_i Y_i\right)^2$$

where Y_i = sum of all observations (y_{ij}s) in ith class
S_e = error sum of squares = $S_T - S_\beta$
V_e = error variance = $S_e/(km - 1)$

The S/N ratio corresponding to the qth run of an OA is given by

$$S/N \text{ ratio} = \eta_q = 10 \log\left(\frac{1}{r}\frac{S_\beta - V_e}{V_e}\right) \tag{4.6}$$

Using Equations 4.1, 4.3, and 4.6, average response tables are constructed to determine useful variables.

Equations for S/N Ratios (MTGS Method)

The following equations are useful if S/N ratios are computed for Gram–Schmidt variables without using OAs.

The Larger-the-Better Type. Let there be t abnormal conditions. Let $u_{j1}, u_{j2},...,u_{jt}$ be the elements of the jth Gram–Schmidt vector (here $j = 1$ to k, if there are k variables). Let s_j be the SD of this vector. Then the S/N ratio (for larger-the-better criterion) corresponding to this variable is

$$\eta_i = -10 \log \left[\frac{1}{t} \sum_{i=1}^{t} \frac{1}{(u_{ji}/s_j)^2} \right] \tag{4.7}$$

The Dynamic Type. As in MTS, in MTGS there are two cases:

1. The true levels of the severity of all the abnormals are known.
2. The true levels of the severity are not known and working averages are used.

CASE 1: THE TRUE LEVELS OF THE SEVERITY OF ALL THE AB-
NORMALS ARE KNOWN.   Let there be t abnormal conditions. Let $u_{j1}, u_{j2},...,u_{jt}$ be the elements of the jth Gram–Schmidt variable. If $M_1, M_2,...,M_t$ represent the true levels of severity (input signals) corresponding to t abnormals, the relationship between the input signal (M_i's) and the output response (y's) is given by

$$y_i = \beta M_i \tag{4.8}$$

where $y_i = \sqrt{u_{ji}^2/s_j^2}$, $i = 1,...,t$, and β is the slope.

The computational procedure for S/N ratios is described as follows. We need to construct an ANOVA table for all Gram–Schmidt variables:

Source	Degrees of Freedom	Sum of Squares
β	1	S_β
e	$t-1$	S_e
Total	T	S_T

where

S_T = total sum of squares = $\Sigma_{i=1}^{t} y_i^2$

r = sum of squares due to input signal = $\Sigma_{i=1}^{t} M_i^2$

S_β = sum of squares due to slope = $(1/r)(\Sigma_{i=1}^{t} M_i y_i)^2$

S_e = error sum of squares = $S_T - S_\beta$

V_e = error variance = $S_e/(t-1)$

The S/N ratio corresponding qth variable is given by

$$S/N \text{ ratio} = \eta_q = 10 \log \left(\frac{1}{r} \frac{S_\beta - V_e}{V_e} \right) \tag{4.9}$$

The estimate of the slope β for the qth variable is given by

$$\beta_q^2 = \frac{1}{r} (S_\beta - V_e) \tag{4.10}$$

CASE 2: DYNAMIC S/N RATIOS WITH WORKING AVERAGES. In some cases, if the true levels of abnormals are not known, it is better to compute dynamic S/N ratios using working averages. Let there be k classes of abnormals with different degrees of severity. The relationship between the input signal (M_is) and output response (ys) is given by

$$y_i = \beta M_i \tag{4.11}$$

where $y_i = \sqrt{u_{ij}^2 / s_i^2}$, $i = 1,...,k$, and β is the slope
    M_i = working mean of the ith class

In Equation 4.11 the slope β is unity. If each class has m observations, then $M_i = (1/m)(\Sigma_{j=1}^{m} y_{ij})$, where $y_{ij} = \sqrt{u_{ij}^2 / s_i^2}$, $i = 1,...,k$, and $j = 1,...,m$.

The computational procedure for S/N ratios is described as

$$S_T = \text{total sum of squares} = \sum_{i=1}^{k} \sum_{j=1}^{m} y_{ij}^2$$

$$r = \text{sum of squares due to input signal} = m \left(\sum_{i=1}^{k} M_i^2 \right)$$

$$S_\beta = \text{sum of squares due to slope} = \frac{I}{r} \left(\sum_{i=1}^{k} M_i Y_i \right)^2$$

where Y_i = sum of all observations (y_{ij}s) in ith class
    S_e = error sum of squares = $S_T - S_\beta$
    V_e = error variance = $S_e/(km - 1)$

The *S/N* ratio corresponding to the qth variable is given by

$$S/N \text{ ratio} = \eta_q = 10 \log \left(\frac{1}{r} \frac{S_\beta - V_e}{V_e} \right) \qquad (4.12)$$

Note. Recently we found that in nondynamic cases, "nominal-the-best" type *S/N* ratios also work well. Since these results were obtained recently, we were unable to publish them in this book.

4.3   ADVANTAGES OF *S/N* RATIOS

In this section, a discussion on the advantages of *S/N* ratios is presented with the help of medical diagnosis data.

4.3.1   *S/N* Ratio as a Simple Measure to Identify Useful Variables

Dr. Kanetaka's data on liver disease testing is used to explain this feature. This data set is the same as that used in Chapter 2. The data contains observations of the healthy group and abnormals on 17 variables, as shown in Table 4.2 (which is same as Table 2.1). The healthy group (MS) is constructed based on observations of 200 people who do not have any health problems. There are 17 abnormal conditions. In this example by mere coincidence the number of variables is equal to the number of abnormal conditions, but this need not be so. The data corresponding to the variables in Table 4.2 are analyzed separately using larger-the-better type *S/N* ratios and dynamic type *S/N* ratios.

The Larger-the-Better-Type S/N Ratio

Larger-the-better type *S/N* ratios are calculated for the data corresponding to variables in Table 4.1 using MTS and MTGS methods. The results of the MTGS are given in Table 4.3. The calculations of *S/N* ratios are based on 17 abnormal conditions. In Table 4.3, $U_1, U_2, ..., U_{17}$ correspond to Gram–Schmidt variables. From this table, it is clear that the variables U_2, U_5, U_6, U_7, U_{10}, U_{12}, U_{13}, U_{14} and U_{15} have higher *S/N* ratios (as compared to other variables), hence they are useful. In the case of MTGS,

TABLE 4.2   Variables in Medical Diagnosis Data

S. No.	Variables	Notation	Notation for Analysis
1	Age		X_1
2	Sex		X_2
3	Total protein in blood	TP	X_3
4	Albumin in blood	Alb	X_4
5	Cholinesterase	ChE	X_5
6	Glutamate O transaminase	GOT	X_6
7	Glutamate P transaminase	GPT	X_7
8	Lactate dehydrogenase	LHD	X_8
9	Alkanline phosphatase	Alp	X_9
10	r-Glutamyl transpeptidase	r-GPT	X_{10}
11	Leucine aminopeptidase	LAP	X_{11}
12	Total cholesterol	TCh	X_{12}
13	Triglyceride	TG	X_{13}
14	Phospholopid	PL	X_{14}
15	Creatinine	Cr	X_{15}
16	Blood urea nitrogen	BUN	X_{16}
17	Uric acid	UA	X_{17}

TABLE 4.3 S/N Ratios (in dB) by the MTGS Method (GSP vectors for 8 abnormal conditions are shown)

	1	2	3	4	5	6	7	8	S/N Ratio
U_1	1.19	1.58	0.71	0.80	0.71	0.61	0.61	1.48	-6.74
U_2	-1.01	-0.89	-1.15	0.97	0.94	0.91	0.91	1.17	0.87
U_3	1.86	-1.83	-1.72	3.09	0.65	-0.07	-0.07	0.17	-13.89
U_4	-0.75	-1.10	-0.84	-1.96	-1.17	0.17	0.17	-1.34	-5.31
U_5	-3.29	-3.82	-3.47	-3.47	-4.10	-3.48	-3.48	-4.08	14.15
U_6	1.59	3.27	0.90	1.27	-0.32	3.99	3.99	0.44	-1.91
U_7	0.47	1.34	-0.30	0.80	0.29	5.45	5.45	1.32	0.58
U_8	3.25	0.72	3.06	2.78	-1.12	2.05	2.05	1.98	-5.24
U_9	0.09	2.72	0.32	-0.14	2.41	1.05	1.05	-0.89	-10.30
U_{10}	-0.24	1.87	0.67	0.44	4.84	-1.27	-1.27	0.57	1.18
U_{11}	-0.49	0.25	0.09	-1.31	-0.29	-0.91	-0.91	-0.20	-18.91
U_{12}	4.58	0.76	0.78	2.96	-0.49	2.85	2.85	3.68	3.75
U_{13}	5.29	0.65	4.17	5.44	1.71	0.35	0.35	8.76	-2.95
U_{14}	0.55	2.03	1.73	0.55	2.06	1.14	2.04	2.02	12.99
U_{15}	1.33	-0.34	3.20	-0.10	-0.59	0.44	0.28	-0.24	-2.83
U_{16}	1.04	-0.71	-1.16	-0.20	-0.97	-1.44	-1.78	0.57	-5.62
U_{17}	0.00	0.12	0.19	-0.71	0.20	0.18	0.18	0.24	-53.45
MD	7.73	8.42	10.29	7.20	10.59	10.56	13.32	14.81	

TABLE 4.4 Variable Allocation in an L_{32} (2^{31}) Array

L_{32} Array Run	X_1 1	X_2 2	X_3 3	X_4 4	X_5 5	X_6 6	X_7 7	X_8 8	X_9 9	X_{10} 10	X_{11} 11	X_{12} 12	X_{13} 13	X_{14} 14	X_{16} 15	X_{16} 16	X_{17} 17	18	19	20	21	22	23	24	25	26	27	28	29	30	31
1	1	1	1	1	1	1	1	1	1	1	1	1	1	1	1	1	1	1	1	1	1	1	1	1	1	1	1	1	1	1	1
2	1	1	1	1	1	1	1	1	1	1	1	1	1	1	1	2	2	2	2	2	2	2	2	2	2	2	2	2	2	2	2
3	1	1	1	1	1	1	1	2	2	2	2	2	2	2	2	1	1	1	1	1	1	1	1	2	2	2	2	2	2	2	2
4	1	1	1	1	1	1	1	2	2	2	2	2	2	2	2	2	2	2	2	2	2	2	2	1	1	1	1	1	1	1	1
5	1	1	1	2	2	2	2	1	1	1	1	2	2	2	2	1	1	1	1	2	2	2	2	1	1	1	1	2	2	2	2
6	1	1	1	2	2	2	2	1	1	1	1	2	2	2	2	2	2	2	2	1	1	1	1	2	2	2	2	1	1	1	1
7	1	1	1	2	2	2	2	2	2	2	2	1	1	1	1	1	1	1	1	2	2	2	2	2	2	2	2	1	1	1	1
8	1	1	1	2	2	2	2	2	2	2	2	1	1	1	1	2	2	2	2	1	1	1	1	1	1	1	1	2	2	2	2
9	1	2	2	1	1	2	2	1	1	2	2	1	1	2	2	1	1	2	2	1	1	2	2	1	1	2	2	1	1	2	2
10	1	2	2	1	1	2	2	1	1	2	2	1	1	2	2	2	2	1	1	2	2	1	1	2	2	1	1	2	2	1	1
11	1	2	2	1	1	2	2	2	2	1	1	2	2	1	1	1	1	2	2	1	1	2	2	2	2	1	1	2	2	1	1
12	1	2	2	1	1	2	2	2	2	1	1	2	2	1	1	2	2	1	1	2	2	1	1	1	1	2	2	1	1	2	2
13	1	2	2	2	2	1	1	1	1	2	2	2	2	1	1	1	1	2	2	2	2	1	1	1	1	2	2	2	2	1	1
14	1	2	2	2	2	1	1	1	1	2	2	2	2	1	1	2	2	1	1	1	1	2	2	2	2	1	1	1	1	2	2
15	1	2	2	2	2	1	1	2	2	1	1	1	1	2	2	1	1	2	2	2	2	1	1	2	2	1	1	1	1	2	2
16	1	2	2	2	2	1	1	2	2	1	1	1	1	2	2	2	2	1	1	1	1	2	2	1	1	2	2	2	2	1	1
17	2	1	2	1	2	1	2	1	2	1	2	1	2	1	2	1	2	1	2	1	2	1	2	1	2	1	2	1	2	1	2
18	2	1	2	1	2	1	2	1	2	1	2	1	2	1	2	2	1	2	1	2	1	2	1	2	1	2	1	2	1	2	1
19	2	1	2	1	2	1	2	2	1	2	1	2	1	2	1	1	2	1	2	1	2	1	2	2	1	2	1	2	1	2	1
20	2	1	2	1	2	1	2	2	1	2	1	2	1	2	1	2	1	2	1	2	1	2	1	1	2	1	2	1	2	1	2
21	2	1	2	2	1	2	1	1	2	1	2	2	1	2	1	1	2	1	2	2	1	2	1	1	2	1	2	2	1	2	1
22	2	1	2	2	1	2	1	1	2	1	2	2	1	2	1	2	1	2	1	1	2	1	2	2	1	2	1	1	2	1	2
23	2	1	2	2	1	2	1	2	1	2	1	1	2	1	2	1	2	1	2	2	1	2	1	2	1	2	1	1	2	1	2
24	2	1	2	2	1	2	1	2	1	2	1	1	2	1	2	2	1	2	1	1	2	1	2	1	2	1	2	2	1	2	1
25	2	2	1	1	2	2	1	1	2	2	1	1	2	2	1	1	2	2	1	1	2	2	1	1	2	2	1	1	2	2	1
26	2	2	1	1	2	2	1	1	2	2	1	1	2	2	1	2	1	1	2	2	1	1	2	2	1	1	2	2	1	1	2
27	2	2	1	1	2	2	1	2	1	1	2	2	1	1	2	1	2	2	1	1	2	2	1	2	1	1	2	2	1	1	2
28	2	2	1	1	2	2	1	2	1	1	2	2	1	1	2	2	1	1	2	2	1	1	2	1	2	2	1	1	2	2	1
29	2	2	1	2	1	1	2	1	2	2	1	2	1	1	2	1	2	2	1	2	1	1	2	1	2	2	1	2	1	1	2
30	2	2	1	2	1	1	2	1	2	2	1	2	1	1	2	2	1	1	2	1	2	2	1	2	1	1	2	1	2	2	1
31	2	2	1	2	1	1	2	2	1	1	2	1	2	2	1	1	2	2	1	2	1	1	2	2	1	1	2	1	2	2	1
32	2	2	1	2	1	1	2	2	1	1	2	1	2	2	1	2	1	1	2	1	2	2	1	1	2	2	1	2	1	1	2

partial correlations are not significant and therefore S/N ratios of Gram–Schmidt variables are computed.

In the MTS method, since there are 17 variables, an L_{32} (2^{31}) orthogonal array is chosen to compute S/N ratios for larger-the-better criterion. The 17 variables are allocated to the first 17 columns of the array, as shown in Table 4.4. For each run or combination abnormal MDs are computed. Since there are 17 abnormals, for each combination there are 17 MDs. Sample results of MTS analysis are given in Table 4.5. The average responses (in terms of S/N ratios) of all the variables are shown in Table 4.6. These responses are calculated for two levels (1's and 2's) of variables allocated to different columns of OA. For example, average response of X_1 at level 1 is the ratio of the sum of S/N ratios corresponding to 1's of X_1 (column 1) to the total number of 1's.

The gain in S/N ratio for a variable is the difference between the average responses of that variable at level 1 and level 2. A positive gain signifies the importance of the variable. From Table 4.6, it is clear that X_2, X_5, X_7, X_{10}, X_{12}, X_{13}, X_{14}, and X_{15} have positive gains and hence they are considered useful variables. Both methods, though different, gave almost the same useful variable combinations: X_2-X_5-X_6-X_7-X_{10}-X_{12}-X_{13}-X_{14}-X_{15} by MTGS and X_2-X_5-X_7-X_{10}-X_{12}-X_{13}-X_{14}-X_{15} by MTS.

The Dynamic-Type S/N Ratio

For the same data, dynamic S/N ratio analysis is carried out for both methods, but only dynamic S/N ratio analysis using MTS method is presented because the same logic can be extended to the MTGS method. As mentioned before, dynamic S/N ratios are used when there are conditions with known levels of severity. From the seventeen abnormal conditions of the data, it was found that the first ten conditions belong to the mild level of severity (mild group) and the remaining seven belong to the medium level of severity (medium group). Therefore, the average of the square root of MDs in these groups is considered as different levels (M_1 and M_2) of input signal (M). The values of these levels are $M_1 = 3.38$ and $M_2 = 6.91$.

The results of S/N ratio analysis using L_{32} array are given in Tables 4.7 and 4.8. From Table 4.8 it is clear that the variables

TABLE 4.5 Larger-the-Better S/N Ratios (in dB) by the MTS Method (MDs of 8 abnormal conditions are shown)

Run	1	2	3	4	5	6	7	8	S/N Ratio
1	7.73	8.42	10.29	7.21	10.59	10.56	13.32	14.81	11.52
2	8.67	9.49	11.54	8.04	11.92	11.78	14.82	16.75	12.01
3	4.34	6.33	4.06	5.08	4.70	14.74	14.74	5.56	8.23
4	4.62	7.99	4.47	6.52	5.40	18.83	18.83	6.40	8.89
5	2.23	3.37	1.49	2.61	6.26	0.66	0.66	0.69	3.19
6	2.36	4.27	1.83	3.10	7.75	0.59	0.59	0.77	3.25
7	7.50	7.93	12.79	6.87	9.28	3.47	8.14	17.37	9.83
8	8.68	10.01	16.11	7.60	11.29	4.22	10.04	21.73	10.55
9	11.77	5.85	7.61	9.30	5.92	4.92	4.92	18.72	9.51
10	13.91	6.26	8.64	11.65	6.35	4.05	4.05	22.31	9.61
11	8.06	7.49	9.50	4.95	9.37	6.64	7.79	8.16	10.23
12	10.31	9.58	11.90	6.20	11.58	7.95	9.44	9.93	11.14
13	4.49	3.32	4.26	2.67	2.40	11.60	12.33	3.88	6.80
14	5.12	3.98	3.50	3.06	2.94	14.45	15.32	4.91	7.43
15	7.17	1.87	1.90	6.50	5.80	11.95	11.95	12.85	8.70
16	8.46	1.34	2.28	7.77	6.03	15.35	15.35	15.66	8.76
17	4.60	8.94	6.58	2.92	15.27	5.12	10.40	9.83	10.27
18	4.49	9.02	6.66	3.10	15.15	4.81	9.91	9.83	10.29
19	8.39	2.80	9.55	6.62	2.96	2.52	2.52	14.42	6.69
20	7.87	2.81	9.66	7.01	2.93	2.41	2.41	13.94	7.01
21	14.01	7.74	15.28	11.37	12.14	15.73	15.73	21.61	12.36
22	13.80	7.74	15.24	11.50	12.01	15.66	15.66	21.37	12.36
23	7.97	16.32	10.01	6.66	16.56	18.30	23.47	16.12	11.96
24	8.01	16.21	10.31	6.46	15.75	17.96	22.91	16.07	11.96
25	5.30	2.63	4.28	4.15	1.98	11.57	11.57	2.25	6.50
26	5.50	2.74	4.96	4.23	1.96	11.60	11.60	2.25	6.87
27	8.20	4.27	2.91	7.17	5.79	12.58	13.51	14.51	9.76
28	7.50	4.63	2.87	7.27	6.06	12.40	13.31	14.09	9.95
29	12.84	7.28	7.00	10.87	5.83	5.62	6.46	19.95	9.93
30	12.88	7.72	8.01	11.12	6.54	7.59	8.52	20.58	10.52
31	8.97	8.37	9.35	5.48	12.09	7.59	7.59	6.13	10.79
32	8.87	8.39	9.30	5.47	12.06	7.64	7.64	6.54	10.83

Note. Level 1, variable is present; level 2, variable is not present.

TABLE 4.6 Average Responses for the Larger-the-Better S/N Ratios

	X_1	X_2	X_3	X_4	X_5	X_6	X_7	X_8	X_9	X_{10}	X_{11}	X_{12}	X_{13}	X_{14}	X_{15}	X_{16}	X_{17}
Level 1	8.73	9.40	8.91	9.28	10.74	9.29	9.63	8.90	8.89	9.71	8.76	10.00	9.94	10.26	9.56	9.14	9.24
Level 2	9.88	9.21	9.69	9.33	7.87	9.31	8.98	9.71	9.71	8.89	9.84	8.61	8.66	8.35	9.05	9.46	9.37
Gain	−1.15	0.19	−0.78	−0.05	2.88	−0.02	0.65	−0.80	−0.82	0.82	−1.08	1.39	1.28	1.91	0.51	−0.32	−0.13

TABLE 4.7   Dynamic *S/N* Ratios (in dB) by the MTS Method (Sample Values)

Run	1	2	3	4	5	6	7	8	*S/N* Ratio
1	7.73	8.42	10.29	7.21	10.59	10.56	13.32	14.81	−6.25
2	8.67	9.49	11.54	8.04	11.92	11.78	14.82	16.75	−6.12
3	4.34	6.33	4.06	5.08	4.70	14.74	14.74	5.56	−10.02
4	4.62	7.99	4.47	6.52	5.40	18.83	18.83	6.40	−10.18
5	2.23	3.37	1.49	2.61	6.26	0.66	0.66	0.69	−10.35
6	2.36	4.27	1.83	3.10	7.75	0.59	0.59	0.77	−10.50
7	7.50	7.93	12.79	6.87	9.28	3.47	8.14	17.37	−7.93
8	8.68	10.01	16.11	7.60	11.29	4.22	10.04	21.73	−8.18
9	11.77	5.85	7.61	9.30	5.92	4.92	4.92	18.72	−9.23
10	13.91	6.26	8.64	11.65	6.35	4.05	4.05	22.31	−9.63
11	8.06	7.49	9.50	4.95	9.37	6.64	7.79	8.16	−3.34
12	10.31	9.58	11.90	6.20	11.58	7.95	9.44	9.93	−3.41
13	4.49	3.32	4.26	2.67	2.40	11.60	12.33	3.88	−10.93
14	5.12	3.98	3.50	3.06	2.94	14.45	15.32	4.91	−11.12
15	7.17	1.87	1.90	6.50	5.80	11.95	11.95	12.85	−6.50
16	8.46	1.34	2.28	7.77	6.03	15.35	15.35	15.66	−7.27
17	4.60	8.94	6.58	2.92	15.27	5.12	10.40	9.83	−7.90
18	4.49	9.02	6.66	3.10	15.15	4.81	9.91	9.83	−7.67
19	8.39	2.80	9.55	6.62	2.96	2.52	2.52	14.42	−10.16
20	7.87	2.81	9.66	7.01	2.93	2.41	2.41	13.94	−9.90
21	14.01	7.74	15.28	11.37	12.14	15.73	15.73	21.61	−5.43
22	13.80	7.74	15.24	11.50	12.01	15.66	15.66	21.37	−5.31
23	7.97	16.32	10.01	6.66	16.56	18.30	23.47	16.12	−7.60
24	8.01	16.21	10.31	6.46	15.75	17.96	22.91	16.07	−7.50
25	5.30	2.63	4.28	4.15	1.98	11.57	11.57	2.25	−11.41
26	5.50	2.74	4.96	4.23	1.96	11.60	11.60	2.25	−11.10
27	8.20	4.27	2.91	7.17	5.79	12.58	13.51	14.51	−5.87
28	7.50	4.63	2.87	7.27	6.06	12.40	13.31	14.09	−4.99
29	12.84	7.28	7.00	10.87	5.83	5.62	6.46	19.95	−9.24
30	12.88	7.72	8.01	11.12	6.54	7.59	8.52	20.58	−8.99
31	8.97	8.37	9.35	5.48	12.09	7.59	7.59	6.13	−5.54
32	8.87	8.39	9.30	5.47	12.06	7.64	7.64	6.54	−5.30

X_4, X_5, X_{10}, X_{12}, X_{13}, X_{14}, X_{15}, and X_{17} have positive gains (when they are part of the system). Hence, these variables are considered useful for future diagnosis. From these discussions, it is clear that by using *S/N* ratios we can identify the useful set of original variables in a simple way.

TABLE 4.8 Average Responses for Dynamic S/N Ratios

	X_1	X_2	X_3	X_4	X_5	X_6	X_7	X_8	X_9	X_{10}	X_{11}	X_{12}	X_{13}	X_{14}	X_{15}	X_{16}	X_{17}
Level 1	-8.185	-8.187	-8.249	-7.949	-7.069	-8.318	-7.976	-8.824	-8.188	-6.358	-8.101	-7.821	-7.562	-7.315	-7.590	-7.982	-7.832
Level 2	-7.745	-7.742	-7.680	-7.980	-8.860	-7.611	-7.954	-7.105	-7.742	-9.571	-7.828	-8.108	-8.367	-8.615	-8.339	-7.947	-8.097
Gain	-0.440	-0.445	-0.568	0.032	1.791	-0.706	-0.022	-1.718	-0.446	3.212	-0.273	0.288	0.804	1.300	0.749	-0.035	0.265

4.3.2   *S/N* Ratio as a Measure of Functionality of the System

The functionality of the system can be expressed in terms of *S/N* ratios. The gain in *S/N* ratio (in dB) indicates the improvement in the functionality of the system. This feature is also explained for both types of *S/N* ratios using Dr. Kanetaka's data.

Gain in S/N Ratio: The Larger-the-Better Type

Since the combination X_2-X_5-X_7-X_{10}-X_{12}-X_{13}-X_{14}-X_{15} is found to be useful in MTS and MTGS methods, the gain in *S/N* ratio is estimated using this combination. From the confirmation run for this combination, the *S/N* ratio is found to be 13.98 dB. The details of the *S/N* ratio analysis are given in Table 4.9. The gain of 2.46 dB in *S/N* ratio indicates that the performance of the system is much better after discarding the variables that are not useful.

Gain in S/N Ratio: Dynamic Type

In the same way as explained in the case of larger-the-better-type analysis, the gain in *S/N* ratio is calculated for dynamic-type analysis. In the case of dynamic-type analysis, the optimal combination is X_4-X_5-X_{10}-X_{12}-X_{13}-X_{14}-X_{15}-X_{17}. The *S/N* ratio details are given in Table 4.10. Here also the gain of 1.98 dB indicates that the performance is better after optimization. For the optimal combination (dynamic type), the estimate of the slope β is given by $\beta^2 = (1/r)(S_\beta - V_e) = 1.313$ and $\beta = 1.14$, which is very close to 1.0, because we are using working averages.

Relationship between Gain in the S/N Ratio and the Variability Reduction

The relationship between gain in the *S/N* ratio and the variability reduction can be given in the form of the following equation:

TABLE 4.9   *S/N* Ratio Analysis (Larger-the-Better Type)

S/N ratio, optimal system	13.98 dB
S/N ratio, original system	11.52 dB
Gain	2.46 dB

TABLE 4.10 *S/N* Ratio Analysis (Dynamic Type)

S/N ratio, optimal system	−4.26 dB
S/N ratio, original system	−6.24 dB
Gain	1.98 dB

$$\text{Variability range}_{\text{improved}} = \left(\frac{1}{2}\right)^{(\text{Gain}/6)} (\text{Variability range}_{\text{initial}})$$

$$(4.13)$$

The reduction in the variability range can also be obtained from the Figure 4.1. In Table 4.11, the reduction in variability range corresponding to the gain in *S/N* ratio of the system (for both types) is given. Thus, based on *S/N* ratios we can estimate the reduction in the variability in predictions using a multidimensional system. In other words, *S/N* ratios can be used to measure the performance of the entire system.

4.3.3 *S/N* Ratio to Predict the Given Conditions

S/N ratios can be effectively used to predict the useful variable combination for the given conditions. This can be accomplished using orthogonal arrays and average response analysis. This method can be useful in the following applications:

- To determine the useful variables for different diseases in medical diagnosis. This will help the doctor concentrate on a

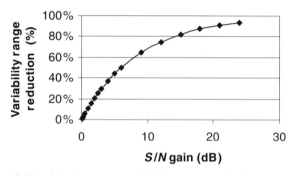

Figure 4.1 *S/N* ratio gain versus variability range reduction.

TABLE 4.11 Reduction in Variability Range Corresponding to the *S/N* Ratio Gain

S/N Ratio Type	Gain	Variability Range Reduction (%)
Larger-the-better	2.46	24.77
Dynamic	1.98	20.47

particular set of variables for a given disease and monitor the patient's condition. In this way, the efficiency of the diagnosis process will be improved and testing time will be reduced.

- To determine the set of variables that are necessary for selling a product in different regions. This will help organizations develop good strategies for different regions.
- To determine the useful variables for setting up laboratories in different regions.
- To identify the useful variables for admitting students into different disciplines of a university.

4.4 CONCLUSIONS

- *S/N* ratio is a very important measure to determine functionality of a multidimensional system.
- In multivariate applications larger-the-better type and dynamic-type *S/N* ratios can be used, though it is always better to use the dynamic-type *S/N* ratios.
- *S/N* ratios are computed using conditions outside MS only, because it is necessary to test the accuracy of the measurement scale based on these conditions.
- This simple measure helps in dimensionality reduction in terms of the original variables.
- The effectiveness of the functionality of the system can be expressed in terms of *S/N* ratio. A gain in *S/N* ratio is a measure of improvement in the effectiveness of the diagnosis system.
- *S/N* ratios can also be used to predict given conditions. This helps in developing a good marketing/sales strategy for a product in different regions. Thus, *S/N* ratio can also be used as a strategic decision-making tool.

5

TREATMENT OF CATEGORICAL DATA IN MTS/MTGS METHODS

In this chapter the applicability of MTS/MTGS methods to categorical (qualitative) data is discussed. The steps to follow for using MTS/MTGS methods with categorical data are provided. The procedure is illustrated with a sales and marketing case study.

5.1 MTS/MTGS WITH CATEGORICAL DATA

This section explains the steps required to use MTS/MTGS methods when there are qualitative variables. When a variable has only two levels (such as gender with levels: males and females), we can use one column to represent that variable with 1's for the first level and 2's for the second level. If there are variables having more than two levels, the following procedure is used.

Let the number of levels of a categorical variable be m. This variable is represented with $m - 1$ columns. If an observation from a sample has level 1, then the first column corresponding to this observation is assigned a value 1 and the remaining columns are assigned 0's. If the observation has level 2, then the second column corresponding to this observation is assigned a value 1 and remaining columns are assigned 0's and so on. But if the observation has the level m, then all the $m - 1$ columns are assigned 0's. For the purpose of better understanding of this procedure, the following illustration is provided.

Let there be k variables in a multivariate diagnosis system. Let the second and third variables be qualitative with levels m_2 and m_3, respectively. To apply the MTS/MTGS methods, the data collection is to be done with the help of the Table 5.1. In this case the number of columns is equal to $k + m_1 + m_2 - 4$. The situation is similar to the case where we have $k + m_1 + m_2 - 4$ variables. Using these columns, MDs are computed in a similar way to the situation where we have $k + m_1 + m_2 - 4$ variables.

Steps in MTS/MTGS with Categorical Data

MTS/MTGS can be applied to a multidimensional system with categorical data in four stages. The steps in each stage are listed here.

Stage I: Construction of a Measurement Scale with Mahalanobis Space (Unit Space) as the Reference

- Define the variables that determine the healthiness of a condition.
- Collect the data on all the variables (including qualitative variables) from the healthy group.
- Compute MDs of all observations using the inverse of the correlation matrix or the Gram–Schmidt method.
- Use the zero point and the unit distance as the reference point or base for the measurement scale.

TABLE 5.1 Data Collection Format for Categorical Data

		Sample Number					
X_1	X_2	X_3	X_4	...	...	X_k	MD
	1 2 ... $(m_1 - 1)$	1 2 ... $(m_2 - 1)$					
	↑	↑					
	$(m_1 - 1)$ columns	$(m_2 - 1)$ columns					
Total columns = $k + m_1 + m_2 - 4$							

Stage II: Validation of the Measurement Scale

- Identify the known conditions outside MS.
- Compute the MDs corresponding to these conditions to validate the scale.

Stage III: Identify the Useful Variables (Developing Stage)

- Find the useful set of variables using orthogonal arrays (OAs) and *S/N* ratios.
- In the case of MTGS using categorical data, the usefulness of a variable is determined based on the levels of that variable. If one of the levels of any variable has a high *S/N* ratio, then that variable is considered useful. However, in MTS, the procedure remains the same, as variables are allocated to the different columns of an orthogonal array. In a given combination if a categorical variable is present, then all the levels of that variable are considered and they are allocated to the columns of OA. The MDs are computed based on the variables in that combination. The gains in *S/N* ratio corresponding to the variables are computed to decide if they are useful.

Stage IV: Future Diagnosis with Useful Variables

- Monitor the conditions using the scale, which is developed with the help of the useful set of variables.

5.2 A SALES AND MARKETING APPLICATION

Because of the variety of potential variables that can influence a buying decision, sales and marketing professionals need a reliable way of finding those variables that have the most influence. MTS /MTGS is ideally suited for this purpose. MTS/MTGS methods have been successfully applied to the analysis of a marketing activity. The study had an immediate practical goal: It was intended to improve the commercial success, as measured by the number of attendees, of an American Supplier Institute (ASI) annual symposia. The analysis is done for 1995 ASI symposium data.

5.2.1   Selection of Suitable Variables

Following a series of brainstorming sessions, a set of variables (Table 5.2) was selected to define the "healthiness" of a condition.

5.2.2   Description of the Variables

In Table 5.2, barring F_1 and F_3, all other variables are of qualitative or categorical type. A description of all these variables is given:

Miles (F_1) represent a measure of the distance from the conference venue to the workplace of the attendee of the symposium. To compute the miles, a special software package was purchased, which provides the numbers of miles between any two cities of the United States.

Section number (F_2) is the American Society for Quality (ASQ) section number, which is categorized by region. There are 15 ASQ regions, denoted 01, 02,...15. We have considered one more region for international members, denoted 16. The international region does not include Canada, because it is classified as region 04 by ASQ. Therefore, the variable F_2 has 16 levels.

ASQ count (F_3) is the number of ASQ members in a particular section.

Job title (F_4) gives the description of the attendee's job. The seven levels for this variable are shown in Table 5.3.

Sponsor/host (F_5) indicates whether the attendee is from the sponsor company or from the host organization or some other company. Therefore, this variable has three levels.

TABLE 5.2   Variables Considered for the Study

Notation	Variable
F_1	Miles
F_2	Section number
F_3	ASQ count
F_4	Job title
F_5	Sponsor/host
F_6	Size of company

TABLE 5.3

Level	Title
L_1	Consultant
L_2	Engineer
L_3	Manager
L_4	Sales and marketing
L_5	Academician
L_6	Trainer
L_7	Others

Size of the company (F_6) has two levels: large-scale or small-scale company. If the company is listed as a Fortune 500 company then, it is considered as large-scale company; otherwise it is a small-scale company.

5.2.3 Construction of Mahalanobis Space

With the above set of six variables, Mahalanobis space (MS) is constructed based on the observations in the healthy group. The group that attended the 1995 symposium is considered to be the healthy group. The healthy group has observations corresponding to 302 attendees. Table 5.4 shows the sample data with MDs for the healthy group. The average MD of this group is 0.998.

5.2.4 Validation of Measurement Scale

After constructing the MS or unit space, which gives the reference point for the scale, the next step is to validate the measurement scale by determining the MDs of known conditions outside MS (abnormals). In this case study, those who did not attend the symposium are considered the abnormals. Fifteen abnormal conditions are selected for this purpose. Table 5.5 shows the complete data of the abnormals along with their MDs.

From this table, it is clear that the abnormals have higher MDs, thus ensuring the accuracy of the scale. Figure 5.1 shows that there is a clear distinction between normals and abnormals in terms of MDs. Hence, this scale can be used to measure the severity of different conditions.

TABLE 5.4 Sample Data of the Healthy Group

Serial No.	Miles F_1 X_1	s_1 X_2	s_2 X_3	s_3 X_4	s_4 X_5	s_5 F_2 X_6	s_6 X_7	s_7 X_8	s_8 X_9	s_9 X_{10}	s_{10} X_{11}	s_{11} X_{12}	s_{12} X_{13}	s_{13} X_{14}	s_{14} X_{15}	s_{15} X_{16}	ASQ count F_3 X_{17}	p1(con) X_{18}	p2(eng) F_4 X_{19}	p3(mgt) X_{20}	p4(s&m) X_{21}	p5(sch) X_{22}	p6(train) X_{23}	Sponsor F_5 X_{24}	Host X_{25}	Size F_6 X_{26}	MD
1	1840	0	0	0	0	0	0	1	0	0	0	0	0	0	0	0	8332	1	0	0	0	0	0	0	0	0	1.037
2	1835	0	0	0	0	0	0	1	0	0	0	0	0	0	0	0	8332	0	0	0	0	0	0	1	0	1	0.985
3	2165	0	0	0	0	0	0	1	0	0	0	0	0	0	0	0	8332	0	1	0	0	0	0	0	0	0	0.742
4	2165	0	0	0	0	0	0	1	0	0	0	0	0	0	0	0	8332	0	1	0	0	0	0	0	0	0	0.676
5	2165	0	0	0	0	0	0	1	0	0	0	0	0	0	0	0	8332	0	1	0	0	0	0	0	0	0	0.743
6	2250	0	0	0	0	0	0	1	0	0	0	0	0	0	0	0	8332	0	1	0	0	0	0	0	0	1	0.677
7	2165	0	0	0	0	0	0	1	0	0	0	0	0	0	0	0	8332	0	1	0	0	0	0	1	0	1	0.808
8	2165	0	0	0	0	0	0	1	0	0	0	0	0	0	0	0	8332	0	0	1	0	0	0	1	0	1	0.843
9	2165	0	0	0	0	0	0	1	0	0	0	0	0	0	0	0	8332	0	0	1	0	0	0	0	0	1	0.668
10	2165	0	0	0	0	0	0	1	0	0	0	0	0	0	0	0	8332	0	0	1	0	0	0	0	0	1	0.668
11	2165	0	0	0	0	0	0	1	0	0	0	0	0	0	0	0	8332	0	0	1	0	0	0	0	0	1	0.668
12	2165	0	0	0	0	0	0	1	0	0	0	0	0	0	0	0	8332	0	0	1	0	0	0	0	0	1	0.668
13	2165	0	0	0	0	0	0	1	0	0	0	0	0	0	0	0	8332	0	0	0	0	0	0	0	0	0	0.743
14	2165	0	0	0	0	0	0	1	0	0	0	0	0	0	0	0	8332	0	0	1	0	0	0	0	0	0	0.597
15	2165	0	0	0	0	0	0	1	0	0	0	0	0	0	0	0	8332	0	0	0	0	1	0	0	0	0	1.557
16	2130	0	0	0	0	0	0	1	0	0	0	0	0	0	0	0	8332	0	0	0	0	0	0	0	0	0	0.742

TABLE 5.5 Abnormal Data with MDs

Serial No.	Miles F_1 X_1	s_1 X_2	s_2 X_3	s_3 X_4	s_4 X_5	s_5 F_2 X_6	s_6 X_7	s_7 X_8	s_8 X_9	s_9 X_{10}	s_{10} X_{11}	s_{11} X_{12}	s_{12} X_{13}	s_{13} X_{14}	s_{14} X_{15}	s_{15} X_{16}	ASQ count F_3 X_{17}	p1(con) X_{18}	p2(eng) F_4 X_{19}	p3(mgt) X_{20}	p4(s&m) X_{21}	p5(sch) X_{22}	p6(train) X_{23}	Sponsor F_5 X_{24}	Host X_{25}	Size F_6 X_{26}	MD
1	463	0	0	0	1	0	0	0	0	0	0	0	0	0	0	0	7201	0	0	0	0	0	0	0	0	0	7.76
2	5	0	0	0	0	0	0	0	0	0	1	0	0	0	0	0	10942	0	0	1	0	0	0	0	0	1	8.28
3	1164	0	0	1	0	0	0	0	0	0	0	0	0	1	0	0	7766	0	0	0	0	0	0	0	0	0	4.10
4	509	0	0	0	0	1	0	0	0	0	0	0	0	0	0	0	6305	0	0	0	0	0	0	0	0	0	3.34
5	10	0	0	0	0	0	0	0	0	0	1	0	0	0	0	0	10942	0	0	0	0	0	0	0	0	0	7.50
6	2077	0	0	0	0	0	1	0	0	0	0	0	0	0	0	0	8723	0	0	0	0	0	0	0	0	0	6.10
7	1074	0	0	0	0	0	0	0	0	0	0	0	0	0	1	0	10570	0	0	0	0	0	0	0	0	0	2.41
8	1204	0	0	0	0	0	0	0	0	0	0	0	0	0	1	0	10570	0	0	0	0	0	0	0	0	1	2.43
9	77	0	0	0	0	0	0	0	0	0	1	0	0	0	0	0	10942	0	0	0	0	1	0	0	0	0	5.23
10	430	0	0	0	0	1	0	0	0	0	0	0	0	0	0	0	7201	0	0	0	0	0	0	0	0	1	3.29
11	72	0	0	0	0	0	0	0	0	0	1	0	0	0	0	0	10942	0	0	0	0	0	0	0	0	0	6.30
12	410	0	0	0	0	1	0	0	0	0	0	0	0	0	0	0	7201	0	0	1	0	0	0	0	0	0	3.67
13	1687	0	0	0	0	0	0	1	0	0	0	0	0	0	0	0	8332	0	0	0	0	0	0	0	0	1	7.79
14	283	0	0	0	0	0	0	0	1	0	0	0	0	0	0	0	7934	0	0	0	0	0	0	0	0	0	5.16
15	446	0	0	0	0	0	0	0	0	0	0	1	0	0	0	0	10623	0	0	0	0	0	0	0	0	0	1.43

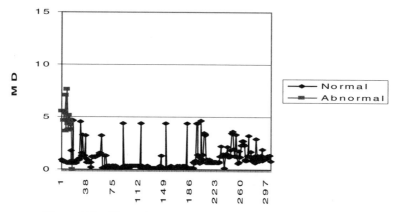

Figure 5.1   Normals versus abnormals with original system.

5.2.5   Identification of the Useful Variables (Developing Stage)

The next stage, after ensuring the accuracy of the measurement scale, is the developing stage. In this stage, a useful set of variables is identified using S/N ratios. To compute S/N ratios different procedures are used in MTS and MTGS. The results of these two methods are discussed in this section.

Results of the MTS Method

To determine the useful variables based on S/N ratios, a suitable orthogonal array (OA) has to be selected to reduce the number of combinations of the variables. In this case the L_8 (2^7) array is selected because there are six variables. The factor allocation in the selected array is shown in the Table 5.6.

In this array for any column, 1 indicates the level that corresponds to presence of the variable and 2 indicates the level that corresponds to absence of the variable. For each combination, three known abnormal conditions are used and S/N ratios are computed based on the MDs of these conditions. Since prior knowledge on the degree of severity is not available, S/N ratios for larger-the-better criterion are used. The MDs along with S/N ratios for all combinations are given in Table 5.7. With S/N ratios, the average responses for the variables are calculated. These results are given in Table 5.8. From the average response table, it

TABLE 5.6   Variable Allocation in L_8 (2^7) Array

			L_8 Array				
Run	F_1	F_2	F_3	F_4	F_5	F_6	
	1	2	3	4	5	6	7
1	1	1	1	1	1	1	1
2	1	1	1	2	2	2	2
3	1	2	2	1	1	2	2
4	1	2	2	2	2	1	1
5	2	1	2	1	2	1	2
6	2	1	2	2	1	2	1
7	2	2	1	1	2	2	1
8	2	2	1	2	1	1	2

is clear that the variables F_1 (miles), F_2 (ASQ section number), F_3 (ASQ count), and F_4 (job title) have positive gains when they are part of the system (level 1) and hence they are useful.

Results of the MTGS Method

The data are also analyzed using the MTGS method. For the purpose of dimensionality reduction, OAs are used, because interpretation of partial correlations might be complex in the cases with categorical data. As we know, if OAs are used, the results would be exactly same as those in MTS.

TABLE 5.7   Abnormal MDs and S/N Ratios for the Combinations of L_8 (2^7) Array

	Factors										
	F_1	F_2	F_3	F_4	F_5	F_6					
Run	1	2	3	4	5	6	7	Abnormal MDs			S/N Ratio
1	1	1	1	1	1	1	1	7.76	4.10	3.67	6.67
2	1	1	1	2	2	2	2	11.06	5.89	4.37	7.88
3	1	2	2	1	1	2	2	0.74	0.67	0.75	−1.43
4	1	2	2	2	2	1	1	0.79	0.65	0.81	−1.30
5	2	1	2	1	2	1	2	3.84	4.79	4.28	6.30
6	2	1	2	2	1	2	1	4.37	5.84	4.37	6.78
7	2	2	1	1	2	2	1	0.91	0.87	0.91	−0.48
8	2	2	1	2	1	1	2	0.56	0.49	0.56	−2.69

TABLE 5.8 **Average Response Table**

	Level 1	Level 2	Gain
F_1	2.95	2.48	0.47
F_2	6.91	−1.47	8.38
F_3	2.85	2.59	0.26
F_4	2.77	2.67	0.10
F_5	2.33	3.10	−0.77
F_6	2.25	3.19	−0.94

Confirmation Run

Since the F_1, F_2, F_3, and F_4 are identified as useful variables, a confirmation run was carried out with these variables. The system with these variables has better discrimination power than the original system. Figure 5.2 shows the distinction between the normals and abnormals with useful variables.

5.2.6 *S/N* Ratio of the System (Before and After)

As we know, the *S/N* ratio can also be used as an effective measure to assess performance of the entire system. Table 5.9 provides the details of *S/N* ratio of the entire system before and after optimization. A gain of 1.154 dB indicates that the system has im-

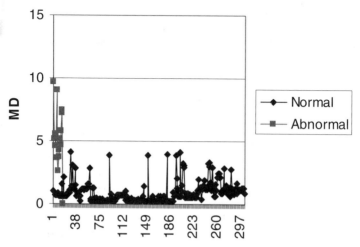

Figure 5.2 Normals versus abnormals with useful variables.

TABLE 5.9 *S/N* **Ratio of the System**

S/N ratio before	6.670
S/N ratio after	7.824
Gain (dB)	1.154

proved significantly after optimization. This improvement is equivalent to a 12.5% reduction in the variability range (from Equation 4.13).

5.3 CONCLUSIONS

- From this case study, it is clear that MTS and MTGS methods can be used in sales and marketing applications.
- The discrimination power of MD appears to be strong, even when working with the categorical data, often found in sales and marketing applications.

6
MTS/MTGS UNDER A NOISE ENVIRONMENT

This chapter discusses the treatment of noise factors in the Mahalanobis–Taguchi system (MTS)/Mahalanobis–Taguchi–Gram–Schmidt process (MTGS). Noise factors are changes in usage environment, such as conditions in different places and the manufacturing variability of the system. Phadke (1989) provided a good discussion on the treatment of noise factors in robust design applications. In this chapter, the different ways of treating noise factors in MTS/MTGS methods are explained with the help of examples.

6.1 MTS/MTGS WITH NOISE FACTORS

In this section, we discuss the treatment of noise factors in MTS/MTGS methods. It is important to know the behavior of a multidimensional system when there are noise factors. The treatment of noise factors can be done in one of the following ways. The question of choosing a particular method of treating a noise factor is a subject of much debate and could be a topic for future research.

1. Treat each level of the noise factor separately and apply MTS/MTGS methods under these levels.

TABLE 6.1   Variables in Medical Diagnosis Data

Sample No.	Variables	Notation	Notation for Analysis
1	Cholinesterase	ChE	X_1
2	Glutamate O transaminase	GOT	X_2
3	Glutamate P transaminase	GPT	X_3
4	Lactate dehydrogenase	LHD	X_4
5	Alkanline phosphatase	Alp	X_5
6	r-Glutamyl transpeptidase	r-GPT	X_6

2. Include the noise factor as one of the variables and determine its usefulness.
3. Combine the variables of all levels of noise factor and apply MTS/MTGS methods with the entire set of variables.
4. Do not consider the noise factor if it cannot be measured.

In the case of a medical diagnosis system, a noise factor could be the place where diagnosis is carried out, like different hospitals. The measurements at different places will induce noise into the system. To demonstrate different ways of treating noise factors, a part of Tatsuji Kanetaka's medical diagnosis data is used. Here only six variables (as shown in Table 6.1) are considered. In this discussion two levels (N_1 and N_2) of a noise factor are considered. The data contains 15 observations corresponding to the healthy people and 5 observations corresponding to the abnormal conditions. The data for the second level is obtained by considering 1 SD (standard deviation) distance from the original data. Since prior information on the levels of severity of abnormal conditions is not known, larger-the-better-type S/N ratios are used to identify the useful set of variables.

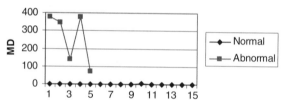

Figure 6.1   Normal versus abnormal conditions with useful set under N_1.

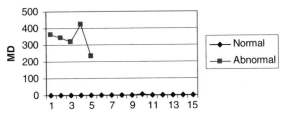

Figure 6.2 Normal versus abnormal conditions with useful set under N_2.

6.1.1 Treat Each Level of the Noise Factor Separately

MTS/MTGS analysis is performed for the two levels of the noise factors (by including all the variables) separately. After ensuring the accuracy of the measurement scale, the L_8 (2^7) orthogonal array (OA) is used to find useful set of variables. The useful variable set for both cases is found to be the same and contains X_2-X_3-X_4-X_5. In both cases, the distinction between the normal group and abnormal conditions is found to be very good, as shown in Figures 6.1 and 6.2.

6.1.2 Include the Noise Factor as One of the Variables

In this case, MTS/MTGS analysis is performed by considering the noise factor as a separate variable. The two levels of the noise factor are used to represent it as variable. Therefore, the number of variables in this case is seven. Since there are seven variables, the L_8 (2^7) OA is used to find a useful set of variables after ensuring accuracy of the scale. Since noise factor is treated as a separate variable, there are 30 observations for the healthy group and 10 observations for abnormal conditions. The arrangement of data is as shown in Table 6.2. In this table, if an observation corresponds to N_1, then $X_7 = 1$; otherwise, $X_7 = 2$. As in the

TABLE 6.2 Arrangement of Data with the Noise Factor as a Variable

Sample No.	X_1	X_2	...	X_6	X_7 (noise)
1					
2					
...					
...					
...					

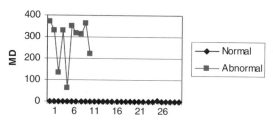

Figure 6.3 Normal versus abnormal conditions with useful set (with noise factor as variable).

previous case, the useful variable set contains X_2-X_3-X_4-X_5. Interestingly, the noise factor is not in the list of useful variables, indicating that it is not important. Figure 6.3 shows the distinction between the normal group and abnormal conditions, which is very good.

6.1.3 Combine Variables of Different Levels of the Noise Factor

In this case, MTS/MTGS analysis is performed by combining the different levels of noise factor. In this example, there are two levels for the noise factor, and the total number of variables to be considered is 12. The arrangement of data is as shown in Table 6.3.

Since there are 12 variables, the L_{16} (2^{15}) OA is used to find the useful set of variables after ensuring the accuracy of the measurement scale. For the first level the useful set of variables is X_1-X_2-X_5-X_6. For the second level this set has X_1-X_3-X_4-X_5-X_6. The measurement scale with the useful variables draws a clear

TABLE 6.3 Arrangement of Data When the Levels of the Noise Factor Are Combined

Sample No.	N_1				N_2				MD
	X_1	X_2	...	X_6	X_1	X_2	...	X_6	
1									
2									
...									
...									

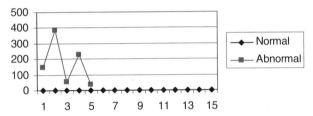

Figure 6.4 Normal versus abnormal conditions with useful set (combining two levels of noise factors).

distinction between the normal group and abnormal conditions, which is shown in Figure 6.4.

6.1.4 Do Not Consider the Noise Factor If It Cannot Be Measured

If the noise factor cannot be measured, MTS/MTGS analysis has to be performed by ignoring the noise factor. This method would not give correct results if the effect of noise is significant. The results of analysis using this procedure are similar to the results of MTS/MTGS analysis at level 1 of Section 6.1.1.

6.2 CONCLUSIONS

- There are different ways of treating the noise factors while using MTS/MTGS methods.
- In all these cases, the measurement scale detects abnormals very well, indicating the power of MD.
- Optimizing the system under different noise conditions would enhance the robustness of a decision-making process.
- The question of choosing a particular method of treating noise factors is a debatable subject and could be a good topic for future research.

7

DETERMINATION OF THRESHOLDS—A LOSS FUNCTION APPROACH

In this chapter, a procedure for determining the threshold in MTS/MTGS methods is discussed. In multivariate systems, determination of a threshold is very important to carrying out the diagnosis process effectively. In this approach, the quadratic loss function (QLF) is used to determine thresholds. In robust engineering, QLF is used to determine the specification limits for a product. Taguchi (1987) provides a clear discussion of QLF. The deviation on either side of the target incurs a monetary loss. This concept helps management understand the importance of robustness of a design, because the variations are explained in monetary terms. It is also recommended that safety factors be set using QLF. The idea of QLF can be extended to determine the thresholds in multivariate diagnosis using MTS/MTGS methods. In classical multivariate methods such as discrimination and classification analysis, classification procedures are developed by minimizing the expected cost of misclassification (ECM). While ECM is minimized, the associated probabilities are considered. The probabilities do not make sense in MTS/MTGS because abnormals are not treated as a separate population.

7.1   WHY THRESHOLD IS REQUIRED IN MTS/MTGS

In MTS/MTGS the threshold can be thought of as the safety factor below which a patient is considered healthy or a manufactured product is acceptable. As we know, MTS/MTGS can be applied to a multidimensional system in the following four stages:

1. Construction of a measurement scale with Mahalanobis space (unit space) as the reference
2. Validation of the measurement scale
3. Identification of the useful variables (developing stage)
4. Future diagnosis with useful variables

The threshold is useful in the fourth stage of MTS/MTGS, because in this stage, while monitoring the conditions, the decision maker (doctor) has to make decisions about conditions. In cases such as medical diagnosis, these four stages have to be applied for each kind of disease (or pattern) after initial diagnosis. Therefore, when MTS/MTGS is applied for diagnosis there are two kinds of thresholds:

General Threshold
   The threshold to determine if the patient requires further examination for a particular disease. For the medical diagnosis case, this number is required to determine the general condition of the patient, i.e., to determine if a patient has some disease.

Specific Threshold
   The second kind of threshold is for a particular disease. For example, in the liver disease case study (discussed in earlier chapters), it is necessary to determine the threshold to know if a particular patient has liver disease.

For the manufacturing inspection example, the general threshold is required for the acceptance of the manufactured product at the end of manufacturing. The specific threshold is required for the acceptance of the product in a given stage of manufacturing.

The threshold has a great impact on the accuracy of the diagnosis process. Incorrect thresholds will result in many false alarms, thereby incurring a huge loss. Therefore, it is necessary to determine the value of threshold in such a way that the total loss is minimized. In this chapter, it is proposed that the quality loss function be used to determine the threshold. In the next section, a brief discussion of QLF is presented.

7.2   QUADRATIC LOSS FUNCTION

In robust engineering, QLF is used to determine the specification limits for a product. The QLF can be used for the following three cases:

1. Nominal-the-best characteristic
2. Larger-the-better characteristic
3. Smaller-the-better characteristic

7.2.1   QLF for the Nominal-the-Best Characteristic

As shown in Figure 7.1, the loss from being away from the target value m is given by $L(y)$. When y is equal to m, the loss is the minimum. The equation for the loss function can be derived as follows:

$$L(y) = L(m + y - m)$$

Using a Taylor series expansion, we have

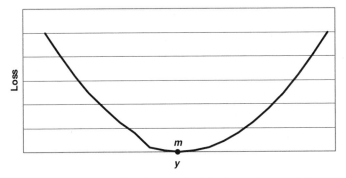

Figure 7.1   QLF for the nominal-the-best characteristic.

$$L(y) = L(m) + \left(\frac{L'(m)}{1!}\right)(y - m) + \left(\frac{L''(m)}{2!}\right)(y - m)^2 + \cdots$$

In general, $L(m) = 0$. Since the loss is minimum at $y = m$, $L'(m) = 0$. Hence, after discarding the higher order terms, $L(y)$ can be written

$$L(y) = k(y - m)^2 \tag{7.1}$$

The proportionality constant in Equation 7.1 can be obtained as follows. If the loss of moving y from m by Λ is equal to $\$A$, then $k = A/\Lambda^2$. Therefore, the QLF in Equation 7.1, can be written

$$L(y) = \frac{A}{\Lambda^2}(y - m)^2 \tag{7.2}$$

7.2.2   QLF for the Larger-the-Better Characteristic

For a larger-the-better characteristic, where the target is infinity, the QLF is shown in Figure 7.2. The equation for the loss function is given by

$$L(y) = k\left(\frac{1}{y^2}\right) \tag{7.3}$$

where $k = A_0 y_0^2$; $A_0 =$ consumer loss and $y_0 =$ consumer tolerance.

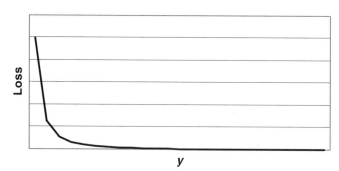

Figure 7.2   QLF for the larger-the-better-type characteristic.

7.2.3 QLF for the Smaller-the-Better Characteristic

For a smaller-the-better characteristic, where the target is zero, the QLF is shown in Figure 7.3. The equation for the loss function is given by

$$L(y) = ky^2 \tag{7.4}$$

where $k = A_0/y_0^2$; $A_0 = $ consumer loss and $y_0 = $ consumer tolerance.

7.3 QLF FOR MTS/MTGS

In this section the usefulness of QLF in the MTS/MTGS methods is explained. The objective of MTS/MTGS is to develop a measurement scale for multidimensional systems. Any error in this scale results in misjudgment. S/N ratios are used to measure the measurement error of the scale. A discussion of the standard error of the scale is provided in Chapter 8. Errors may also occur due to incorrect thresholds used for diagnosis. Because of incorrect thresholds, there could be adverse effects. For example, a patient who does not require treatment might be treated or a patient who requires treatment might be left untreated. Both are very costly mistakes and result in huge loss.

To carry out a good diagnosis, the decision maker must determine the threshold judiciously. In medical applications, the threshold is the balance point between the cost of treating a patient and the cost of not treating a patient. In a student admission system,

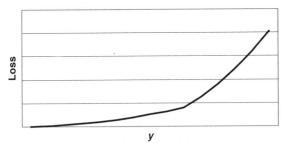

Figure 7.3 QLF for the smaller-the-better-type characteristic.

it is the balance point between the cost of admitting a student and the cost of not admitting a student. The threshold is like the LD_{50} number. A product having this value may be considered a good product or a bad product. Similarly, a patient having the value of the threshold may be considered healthy or unhealthy, which means that 50% of the patients having this value are healthy and the remaining 50% are unhealthy. This number can be obtained by using QLF. The loss function curve can be obtained by plotting the square root of MDs against the corresponding losses. In MTS/MTGS, QLF for the case of smaller-the-better characteristic is applicable since the target value of MD is zero. The shape of the curve will be similar to that of Figure 7.3. The representation of QLF in MTS/MTGS is as in Figure 7.4.

Let Λ_0 be the functional limit. This value corresponds to the distance when a patient dies or when there is a manufacturing product having the functional failure. If A_0 is the loss associated with Λ_0, then the proportionality constant in the QLF equation, $k = A_0/\Lambda_0^2$, and the corresponding loss is given by

$$\text{Loss} = \frac{A_0}{\Lambda_0^2} \text{MD} \tag{7.5}$$

7.3.1 Determination of Threshold

In this section a procedure for determining the threshold in MTS/MTGS is outlined. There are two ways of determining the thresholds.

Procedure 1

In the case of medical diagnosis, if the doctor knows the set of people who require treatment, then the average squared distance (Λ^2) corresponding to these patients can be used as a threshold; i.e., $T = \Lambda^2$. It means that this value is based entirely on the doctor's judgment. This logic can be easily extended to cases other than the medical diagnosis.

Procedure 2

Let Λ_0 be the functional limit, which is the critical distance. For the medical diagnosis case this is the distance at which a patient

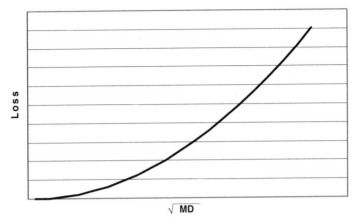

Figure 7.4   QLF for MTS/MTGS.

would die because of illness. Similarly, for the student system this distance corresponds to a student who has not graduated. Let A_0 be the loss associated with the functional limit. Then the loss function equation can be obtained as Loss $= (A_0/\Lambda_0^2)$ MD.

Let us assume that the cost of further diagnosis of a patient is $\$A$. For the student system A is the cost associated with providing additional training to the student who did not graduate. Let Λ be the distance corresponding to A. Then Λ can be obtained by

$$\Lambda = \sqrt{\frac{A}{A_0}}\,\Lambda_0 \tag{7.6}$$

Therefore, if T is the threshold, we can write

$$\Lambda = \sqrt{T} \tag{7.7}$$

or

$$T = \Lambda^2 \tag{7.8}$$

For the case of medical diagnosis, using the threshold T, the decision rule can be stated as follows:

If the value of MD is equal to T, then the patient is healthy or unhealthy. If MD is below T, then the patient is healthy and if MD is above T, then the patient is unhealthy.

7.3.2 When Only Good Abnormals Are Present

When good abnormals are present (as in the student admission system), we can compute the threshold to determine which student can be awarded a scholarship or a fellowship. Good abnormals mean the students are exceptionally good with top scores in all or most of the subjects.

7.4 EXAMPLES

In this section determining the threshold is discussed for the medical diagnosis and the student admissions system cases.

7.4.1 Medical Diagnosis Case

The threshold is calculated for Dr. Kanetaka's liver disease example with 17 variables, as explained in earlier chapters. Since the threshold is calculated for a particular disease, this is a case of specific threshold, which has been explained in Section 7.1. Assume that the functional limit Λ_0 at which a health examinee with liver disease dies is 120. This value is assumed to be double the value of the distance (60) when a subjective symptom is manifest.

The loss of the person (A_0) who died without undergoing close examination is expressed by

$A_0 = $ (Annual income)

$\times$ (Average lifetime expectancy from the date of examination)

$A_0 = $ (six million yen/year)(27 years)

Therefore, the equation for the loss function is

$$\text{Loss} = \left[\frac{(\text{six million yen/year})(27 \text{ years})}{120^2} \right] \text{MD}$$

Computation of the Threshold

If the cost of treatment of a person and the opportunity loss of taking off from the work (A) is 100,000 yen, then the corresponding distance (Λ) can be obtained from Equation 7.6. Substituting the values in this equation, we have $\Lambda = 3$. Therefore, the corresponding threshold $T = \Lambda^2 = 3^2 = 9$. The QLF curve for this example is as shown in Figure 7.5. The square root of the threshold is shown with a solid line.

7.4.2 A Student Admission System

If the functional limit Λ_0 corresponding to a bad student (a student who did not graduate) for a masters degree is 14, then the loss A_0 corresponding to such a student can be obtained as

$A_0 =$ Tuition fee for masters program

+ (Loss of annual income from the time of graduation)

× (Number of years of service from graduation date)

Let the cost of tuition for a graduate study = \$12,000. Let us assume that the average age of a student at the time of graduation

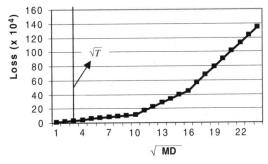

Figure 7.5 QLF for the medical diagnosis example.

is 25 years and the average retirement age of any person is 60 years. Let us further assume that the average level of salary for a master's graduate is $50,000/year and for a bachelors degree holder is $25,000/year. Then we have

$$A_0 = \$12,000 + 35(50,000 - 25,000) = 887,000$$

Therefore, the equation for the loss function is

$$\text{Loss} = \left(\frac{887,000}{14^2}\right) \text{MD}$$

Let the loss for a student to improve the performance (A) be $22,000. The loss A is sum of tuition and the cost of additional training. In that case the corresponding distance Λ can be obtained from Equation 7.6. Substituting the values in this equation we have $\Lambda = 2.20$. Therefore, the corresponding threshold $T = \Lambda^2 = 2.2^2 = 4.84$. As in the case of medical diagnosis, the QLF curve can be drawn for this example.

7.5 CONCLUSIONS

- To carry out effective diagnosis using MTS/MTGS methods, the threshold should be judiciously determined.
- Because of incorrect thresholds, a patient who does not re- quire treatment might be treated or a patient who requires treatment might be left untreated. Both are very costly mis- takes and result in huge loss. Hence, the value of threshold has a significant impact on the measurement scale developed using MTS/MTGS methods.
- The concept of QLF can be extended to the multidimensional systems to determine thresholds, which will enable decision makers to perform diagnosis more accurately.
- The thresholds determined with the help of QLF will help minimize false alarms.

8
STANDARD ERROR OF THE MEASUREMENT SCALE

Application of MTS/MTGS methods can be a subject of measurement science for multidimensional systems. The development of the scale is based on the input characteristics of the system. In medical diagnosis, for example, the aim of such a scale is to measure the degree of severity of each disease based on this scale. For any measurement scale, if the standard error is known, then the scale can be suitably calibrated to obtain accurate measurements. In this chapter, a method of determining standard error of multidimensional measurement scale is discussed.

8.1 WHY MAHALANOBIS DISTANCE IS USED FOR CONSTRUCTING THE MEASUREMENT SCALE

Mahalanobis distance (MD) is used in MTS/MTGS methods because of its ability to distinguish different conditions in multidimensional systems. We must construct the Mahalanobis space (MS) to define a zero point and unit distance, which are used as the base or reference point to the scale. From the zero point the distances of all conditions are measured to find the degree of abnormality. Unit distance is useful in defining the MS, in general, irrespective of the number of variables. In other words, we can say that the zero point and unit distance are the properties of the

MS. In Chapter 2 it was proved that the Mahalanobis space can be constructed using the observations in a healthy group irrespective of the type of input variables and their distributions.

8.2   STANDARD ERROR OF THE MEASUREMENT SCALE

In MTS/MTGS methods, when dynamic S/N ratios are used to find the set of useful variables, there are two types of situations:

1. The true levels of severity are known.
2. The true levels are not known and working averages are used.

If the true levels of the severity are known or if the working averages are used, it is possible to estimate the standard error (σ^2) of the measurement scale. If the true levels are known, the standard error is obtained by calibrating the measurement scale by β^2.

The standard error is given by

$$\sigma^2 = \frac{\sigma_e^{\,2}}{\beta^2} \tag{8.1}$$

where

$\sigma_e^{\,2}$ = variance around the slope = error variance V_e
β = estimate of the slope

From Equation 8.1, it is clear that the standard error is the reciprocal of the S/N ratio. If the working averages are used, then the estimate of the slope is 1.0. Therefore, it is not necessary to calibrate the measurement scale. In this case, the standard error is composed of two parts:

$$\sigma^2 = \sigma_e^{\,2} + \sigma_m^{\,2} \tag{8.2}$$

where

σ_e^2 = error variance (variance of MD from the working mean)
σ_m^2 = variation between the abnormals in each class

Since in this case the true levels of abnormals are not known, it is not possible to separate σ_m^2 from the standard error. However, it will not affect the decision to select the useful variables, because σ_m^2 will be same for all runs of an orthogonal array.

8.3   STANDARD ERROR FOR THE MEDICAL DIAGNOSIS

In this section, computational procedure of the standard error is explained for the liver disease test example, discussed in Chapter 4. There are 17 variables, the healthy group has observations on 200 people, the number of abnormal conditions t is 17, and the levels of the input signal are $M_1 = 3.38$ and $M_2 = 6.91$. Here, dynamic-type S/N ratios with working averages are used. There are two groups: mild group with 10 observations and medium group with 7 observations. The useful-variable combination was found to be X_4-X_5-X_{10}-X_{12}-X_{13}-X_{14}-X_{15}-X_{17}. For this combination,

S_T = total sum of squares = $\displaystyle\sum_{i=1}^{t} D_i^2 = 650.2$ (where D_i^2 is MD

corresponding to ith abnormal condition and $t = 17$)

r = sum of squares due to the input signal = $10\,M_1^2 + 7\,M_2^2$
= 449.8

S_β = sum of squares due to slope = $(M_1 Y_1 + M_2 Y_2)^2 / r = 594.1$

S_e = error sum of squares = $S_T - S_\beta = 56.1$

V_e = error variance = $S_e/(t - 1) = 3.51$

The estimate of the slope β is given by

TABLE 8.1 Comparison between the Original System and the Optimal System

Combination	Standard Error	S/N Ratio (dB)
Original	4.18	−6.24
Optimal	3.51	−4.26
Improvement	0.67	1.98

$$\beta^2 = \frac{1}{r}(S_\beta - V_e) = 1.313$$

and $\beta = 1.14$, which is very close to 1.0. Therefore, the standard error of this scale $= V_e = 3.51$.

The S/N ratio corresponding to this combination is

$$S/N \text{ ratio} = \eta = 10 \log\left[\left(\frac{1}{r}\right)\left(\frac{S_\beta - V_e}{V_e}\right)\right] = -4.26 \text{ dB}$$

It was found that after optimization there is a reduction in the standard error (correspondingly an increase in S/N ratio). Table 8.1 summarizes these results.

8.4 CONCLUSIONS

- It is possible to estimate the standard error of the measurement scale developed using MTS/MTGS methods. The reduction of standard error will help in obtaining accurate measurements from the scale. The procedure for estimation of standard error depends on the prior knowledge of the conditions or working averages.
- S/N ratio, which has several other advantages, can also be used to determine the standard error of the scale in multivariate applications.

9
ADVANCE TOPICS IN MULTIVARIATE DIAGNOSIS

In this chapter advanced topics related to multivariate diagnosis are discussed. These topics are aimed at overcoming difficulties due to multicollinearity (strong correlations), and small correlations, improving the process of subset selection for complex problems, and selection of Mahalanobis space from historical data. If there are incidences of strong correlations between two variables, one method is to discard one variable and retain the other for analysis. However, it is recommended that no variable be discarded until S/N ratio analysis is performed to identify the useful variable set, since in a multidimensional system every variable and the correlation structure are important. In Chapter 3, it was demonstrated that multicollinearity problems can be overcome to some extent, using the MTGS method. In this chapter a superior and more robust method of handling the multicollinearity problems is explained with examples. This method can be used even in cases where there are no multicollinearity problems. In this method, an adjoint matrix is used to calculate distances, instead of an inverse matrix. Another important situation is the case of small correlations. This chapter outlines a methodology to take care of the situations of small correlations by using the β-adjustment method. The chapter also provides a better way of selecting subsets for complex problems where there are several hundred variables. This is done by introducing a new metric called multiple Mahalanobis distance (MMD). Finally, the chapter briefly outlines a procedure for selecting a Mahalanobis space from historical data.

9.1   MULTIVARIATE DIAGNOSIS USING THE ADJOINT MATRIX METHOD

Before going into the details of this method, it is necessary to understand the procedure used to compute inverse matrices, which are used in the MTS method. If A is a square matrix (an inverse can be computed for square matrices only), then its inverse A^{-1} is given as

$$A^{-1} = \frac{1}{\det. A} A_{adj} \tag{9.1}$$

where A_{adj} is called adjoint matrix of A. An adjoint matrix is a transpose of a cofactor matrix, which is obtained by cofactors of all the elements of matrix A. The determinant of matrix A (det. A) is a characteristic number (scalar) associated with a square matrix. A matrix is said to be singular if its determinant is zero.

9.1.1   Related Topics of Matrix Theory

Determinant of a Matrix

As mentioned before, the determinant is a characteristic number associated with a square matrix. The importance of a determinant can be realized when solving a system of linear equations using matrix algebra. The solution to the system of equations contains an inverse matrix term, which is obtained by dividing the adjoint matrix by the determinant. If the determinant is zero, then the solution does not exist.

Consider a 2 × 2 matrix:

$$A = \begin{bmatrix} a_{11} & a_{12} \\ a_{21} & a_{22} \end{bmatrix}$$

The determinant of this matrix is $a_{11} a_{22} - a_{12} a_{21}$. Now consider a 3 × 3 matrix:

$$A = \begin{bmatrix} a_{11} & a_{12} & a_{13} \\ a_{21} & a_{22} & a_{23} \\ a_{31} & a_{32} & a_{33} \end{bmatrix}$$

The determinant of A can be calculated as

$$\det. A = a_{11}A_{11} + a_{12}A_{12} + a_{13}A_{13}$$

where $A_{11} = (a_{22}\,a_{33} - a_{23}a_{32})$, $A_{12} = -(a_{21}\,a_{33} - a_{23}a_{31})$, $A_{13} = (a_{21}a_{32} - a_{22}a_{31})$ are cofactors of the elements a_{11}, a_{12}, and a_{13} of matrix A, respectively. The cofactors can be computed from submatrices obtained by deleting the rows and columns passing through the respective elements. Along a row or a column, the cofactors will have alternate plus and minus signs, with the first cofactor having a positive sign.

The above equation for the determinant is obtained using the elements of the first row and their cofactors. The same value of determinant can be obtained by using other rows or any column of the matrix with corresponding cofactors. In general, the determinant of an $n \times n$ square matrix can be written as

$$\det. A = a_{i1}A_{i1} + a_{i2}A_{i2} + \cdots + a_{in}A_{in}$$

along any row i, where $i = 1, 2, ..., n$

or

$$\det. A = a_{1j}A_{1j} + a_{2j}A_{2j} + \cdots + a_{nj}A_{nj}$$

along any column j, where $j = 1, 2, ..., n$

Cofactor

From the above discussion, it is clear that the cofactor A_{ij} of an element a_{ij} is the factor remaining after the element a_{ij} is factored out. The method of computing the cofactors is explained above for a 3×3 matrix. Along a row or a column the cofactors will have alternate signs of positive and negative, with the first cofactor having a positive sign.

Adjoint Matrix of a Square Matrix

The adjoint of a square matrix A is obtained by replacing each element of A with its own cofactor and transposing the result. Again, consider a 3×3 matrix:

$$A = \begin{bmatrix} a_{11} & a_{12} & a_{13} \\ a_{21} & a_{22} & a_{23} \\ a_{31} & a_{32} & a_{33} \end{bmatrix}$$

The cofactor matrix containing cofactors (A_{ij}'s) of the elements of the above matrix can be written:

$$A = \begin{bmatrix} A_{11} & A_{12} & A_{13} \\ A_{21} & A_{22} & A_{23} \\ A_{31} & A_{32} & A_{33} \end{bmatrix}$$

The adjoint of the matrix A, which is obtained by transposing the cofactor matrix, can be written:

$$\text{Adj. } A = \begin{bmatrix} A_{11} & A_{21} & A_{31} \\ A_{12} & A_{22} & A_{32} \\ A_{13} & A_{23} & A_{33} \end{bmatrix}$$

Inverse Matrix

The inverse of matrix A (denoted as A^{-1}) can be obtained by dividing the elements of the adjoint of A by the determinant.

Singular and Nonsingular Matrices

If the determinant of a square matrix is zero, then it is called a singular matrix. Otherwise, the matrix is known as nonsingular.

9.1.2 Adjoint Matrix Method for Handling Multicollinearity

Multicollinearity problems arise out of strong correlations. When there are strong correlations, the determinant of a correlation matrix will tend to become zero, thereby making the matrix almost singular. In such cases, the inverse matrix will be inaccurate or cannot be computed (because the determinant term is in the denominator of Equation 9.1). As a result scaled MDs will also be inaccurate. Such problems can be avoided if we use a matrix form, which is not affected by the determinant term. From Equation 9.1, it is clear that the adjoint matrix satisfies this requirement.

We know that MDs in the MTS method are computed using inverse of the correlation matrix (C^{-1}, where C is correlation matrix). In the adjoint matrix method, it is proposed that the adjoint matrix be used to calculate the distances. If the MDA denotes the distances obtained from the adjoint matrix method, then the equation for MDA can be written as

$$\text{MDA}_j = \frac{1}{k} Z_{ij}' C_{\text{adj}} Z_{ij} \tag{9.2}$$

where

$j = 1$ to n
$Z_{ij} = (z_{1j}, z_{2j}, ..., z_{kj})$
 $=$ standardized vector obtained by standardized values of X_{ij}
  $(i = 1, ..., k)$
$Z_{ij} = (X_{ij} - m_i)/s_i$
$X_{ij} =$ value of the ith characteristic in the jth observation.
$m_i =$ mean of the ith characteristic
$s_i =$ SD of the ith characteristic
$' =$ transpose of the vector
$C_{\text{adj}} =$ adjoint of the correlation matrix

Equation 9.2 is similar to the Equation 2.2. The relationship between the MD (scaled MD) in Equation 2.2 and the MDAs in (9.2) can be written:

$$\text{MD}_j = \frac{1}{\text{det. } C} \text{MDA}_j \tag{9.3}$$

Thus, MDAs are similar to MDs with different properties; that is, the average MDA is not unity. As in the case of MDs, MDAs represent the distances from the healthy group and can be used to measure the degree of abnormalities. In the adjoint matrix method also, the Mahalanobis space contains means, standard deviations, and correlation structure of the normal or healthy group. Here, the Mahalanobis space cannot be called unit space, since the average of the MDAs is not unity.

9.2   EXAMPLES FOR THE ADJOINT MATRIX METHOD

9.2.1   Example 1

The adjoint matrix method is applied to the medical diagnosis example with the 17 variables discussed in earlier chapters. The correlation matrix, inverse matrix, and adjoint matrix corresponding to the 17 variables are given in Tables 9.1, 9.2, and 9.3, respectively. In this case the determinant of the correlation matrix is 0.00001314. The Mahalanobis distances (sample values), calculated by the inverse matrix method and the adjoint matrix method, are given in Table 9.4 for normals and Table 9.5 for abnormals. From Table 9.4, it is clear that the average MDAs for normals do not converge to 1.0.

As before, the $L_{32}(2^{31})$ OA is used to accommodate 17 variables. Table 9.6 gives dynamic S/N ratios for all the combinations of this array using the inverse matrix method and adjoint matrix method. Table 9.7 shows the gain in S/N ratios for both methods. Gains in S/N ratios are same for both methods. The important variable combination based on these gains is X_4-X_5-X_{10}-X_{12}-X_{13}-X_{14}-X_{15}-X_{17}. Hence, even if we use the adjoint matrix method the ultimate results would be the same. However, the MDA is advantageous because it will not take into account the determinant of the correlation matrix. If the multicollinearity problem exists, the inverse matrix becomes inefficient, giving rise to inaccurate MDs. Such problems can be avoided if MDAs are used based on the adjoint matrix method.

The above example shows the applicability of the adjoint matrix method. The following example illustrates that the adjoint matrix method performs better when there are problems of multicollinearity.

TABLE 9.1 Correlation Matrix

	X_1	X_2	X_3	X_4	X_5	X_6	X_7	X_8	X_9	X_{10}	X_{11}	X_{12}	X_{13}	X_{14}	X_{15}	X_{16}	X_{17}
X_1	1.000	-0.297	-0.278	-0.403	-0.220	0.101	0.041	0.208	0.293	-0.104	-0.112	0.264	0.135	0.283	-0.292	-0.019	-0.282
X_2	-0.297	1.000	0.103	0.416	0.690	0.287	0.379	-0.108	-0.048	0.647	0.395	-0.237	0.269	-0.222	0.886	0.254	0.798
X_3	-0.278	0.103	1.000	0.427	0.202	0.084	0.139	0.072	0.011	0.177	0.182	0.070	0.158	0.078	0.150	-0.119	0.198
X_4	-0.403	0.416	0.427	1.000	0.315	0.038	0.056	0.010	-0.106	0.269	0.219	-0.136	0.100	-0.135	0.365	0.091	0.413
X_5	-0.220	0.690	0.202	0.315	1.000	0.345	0.385	0.063	-0.057	0.578	0.429	0.012	0.367	0.032	0.644	0.135	0.675
X_6	0.101	0.287	0.084	0.038	0.345	1.000	0.790	0.316	0.177	0.550	0.535	0.148	0.320	0.168	0.321	0.181	0.359
X_7	0.041	0.379	0.139	0.056	0.385	0.790	1.000	0.143	0.068	0.568	0.507	0.134	0.373	0.148	0.398	0.109	0.435
X_8	0.208	-0.108	0.072	0.010	0.063	0.316	0.143	1.000	0.229	0.129	0.227	0.250	0.103	0.257	-0.063	0.095	-0.015
X_9	0.293	-0.048	0.011	-0.106	-0.057	0.177	0.068	0.229	1.000	0.065	0.147	0.171	0.121	0.168	-0.075	-0.096	-0.061
X_{10}	-0.104	0.647	0.177	0.269	0.578	0.550	0.568	0.129	0.065	1.000	0.683	0.052	0.437	0.079	0.612	0.138	0.649
X_{11}	-0.112	0.395	0.182	0.219	0.429	0.535	0.507	0.227	0.147	0.683	1.000	0.159	0.342	0.152	0.445	0.048	0.465
X_{12}	0.264	-0.237	0.070	-0.136	0.012	0.148	0.134	0.250	0.171	0.052	0.159	1.000	0.310	0.967	-0.140	-0.004	-0.023
X_{13}	0.135	0.269	0.158	0.100	0.367	0.320	0.373	0.103	0.121	0.437	0.342	0.310	1.000	0.318	0.267	-0.041	0.352
X_{14}	0.283	-0.222	0.078	-0.135	0.032	0.168	0.148	0.257	0.168	0.079	0.152	0.967	0.318	1.000	-0.119	0.025	-0.011
X_{15}	-0.292	0.886	0.150	0.365	0.644	0.321	0.398	-0.063	-0.075	0.612	0.445	-0.140	0.267	-0.119	1.000	0.279	0.771
X_{16}	-0.019	0.254	-0.119	0.091	0.135	0.181	0.109	0.095	-0.096	0.138	0.048	-0.004	-0.041	0.025	0.279	1.000	0.177
X_{17}	-0.282	0.798	0.198	0.413	0.675	0.359	0.435	-0.015	-0.061	0.649	0.465	-0.023	0.352	-0.011	0.771	0.177	1.000

TABLE 9.2 Inverse Matrix

	X_1	X_2	X_3	X_4	X_5	X_6	X_7	X_8	X_9	X_{10}	X_{11}	X_{12}	X_{13}	X_{14}	X_{15}	X_{16}	X_{17}
X_1	1.592	-0.003	0.307	0.297	0.118	-0.082	-0.116	-0.193	-0.304	-0.113	0.248	0.337	-0.284	-0.552	0.146	-0.028	0.198
X_2	-0.003	8.136	0.658	-0.706	-1.281	0.627	-0.439	0.379	-0.576	-1.482	0.748	-0.192	-0.077	1.358	-4.277	-0.316	-1.525
X_3	0.307	0.658	1.442	-0.594	-0.169	0.136	-0.258	-0.066	-0.123	-0.115	0.070	0.223	-0.097	-0.304	-0.315	0.194	-0.023
X_4	0.297	-0.706	-0.594	1.677	0.101	0.009	0.272	-0.143	0.088	0.071	-0.157	0.026	-0.049	0.055	0.317	-0.103	-0.296
X_5	0.118	-1.281	-0.169	0.101	2.357	-0.197	0.110	-0.193	0.200	-0.034	-0.121	0.210	-0.235	-0.440	0.077	0.108	-0.429
X_6	-0.082	0.627	0.136	0.009	-0.197	3.403	-2.266	-0.483	-0.297	-0.436	-0.348	0.332	0.044	-0.156	-0.108	-0.338	-0.104
X_7	-0.116	-0.439	-0.258	0.272	0.110	-2.266	3.192	0.275	0.252	-0.172	-0.133	-0.240	-0.195	0.106	-0.009	0.147	-0.153
X_8	-0.193	0.379	-0.066	-0.143	-0.193	-0.483	0.275	1.338	-0.157	-0.056	-0.179	-0.103	0.044	-0.028	0.022	-0.143	0.012
X_9	-0.304	-0.576	-0.123	0.088	0.200	-0.297	0.252	-0.157	1.247	0.101	-0.218	-0.118	-0.034	-0.006	0.240	0.157	0.131
X_{10}	-0.113	-1.482	-0.115	0.071	-0.034	-0.436	-0.172	-0.056	0.101	3.321	-1.247	0.928	-0.335	-1.004	0.386	0.041	-0.350
X_{11}	0.248	0.748	0.070	-0.157	-0.121	-0.348	-0.133	-0.179	-0.218	-1.247	2.302	-0.880	-0.001	0.754	-0.637	0.151	-0.036
X_{12}	0.337	-0.192	0.223	0.026	0.210	0.332	-0.240	-0.103	-0.118	0.928	-0.880	16.234	-0.293	-15.614	0.589	0.274	-0.363
X_{13}	-0.284	-0.077	-0.097	-0.049	-0.235	0.044	-0.195	0.044	-0.034	-0.335	-0.001	-0.293	1.537	-0.096	0.043	0.167	-0.145
X_{14}	-0.552	1.358	-0.304	0.055	-0.440	-0.156	0.106	-0.028	-0.006	-1.004	0.754	-15.614	-0.096	16.526	-0.826	-0.463	-0.018
X_{15}	0.146	-4.277	-0.315	0.317	0.077	-0.108	-0.009	0.022	0.240	0.386	-0.637	0.589	0.043	-0.826	5.415	-0.330	-0.691
X_{16}	-0.028	-0.316	0.194	-0.103	0.108	-0.338	0.147	-0.143	0.157	0.041	0.151	0.274	0.167	-0.463	-0.330	1.249	0.120
X_{17}	0.198	-1.525	-0.023	-0.296	-0.429	-0.104	-0.153	0.012	0.131	-0.350	-0.036	-0.363	-0.145	-0.018	-0.691	0.120	3.599

TABLE 9.3 Adjoint Matrix

	X_1	X_2	X_3	X_4	X_5	X_6	X_7	X_8	X_9	X_{10}	X_{11}	X_{12}	X_{13}	X_{14}	X_{15}	X_{16}	X_{17}
X_1	2.09E-05	-3.8E-08	4.03E-06	3.9E-06	1.55E-06	-1.07E-06	-1.52E-06	-2.53E-06	-4E-06	-1.49E-06	3.26E-06	4.43E-06	-3.73E-06	-7.25E-06	1.92E-06	-3.63E-07	2.6E-06
X_2	-3.8E-08	0.000107	8.65E-06	-9.27E-06	-1.68E-05	8.24E-06	-5.77E-06	4.98E-06	-7.57E-06	-1.95E-05	9.83E-06	-2.53E-06	-1.01E-06	1.78E-05	-5.62E-05	-4.16E-06	-2E-05
X_3	4.03E-06	8.65E-06	1.89E-05	-7.81E-06	-2.22E-06	1.78E-06	-3.4E-06	-8.65E-07	-1.62E-06	-1.51E-06	9.22E-07	2.93E-06	-1.27E-06	-3.99E-06	-4.13E-06	2.55E-06	-3.04E-07
X_4	3.9E-06	-9.27E-06	-7.81E-06	2.2E-05	1.33E-06	1.18E-06	3.57E-06	-1.88E-06	1.16E-06	9.35E-07	-2.06E-06	3.41E-07	-6.46E-07	7.2E-07	4.17E-06	-1.36E-06	-3.89E-06
X_5	1.55E-06	-1.68E-05	-2.22E-06	1.33E-06	3.1E-05	-2.59E-06	1.44E-06	-2.54E-06	2.63E-06	-4.5E-07	-1.6E-06	2.77E-06	-3.09E-06	-5.78E-06	1.02E-06	1.42E-06	-5.64E-06
X_6	-1.07E-06	8.24E-06	1.78E-06	1.18E-06	-2.59E-06	4.47E-05	-2.98E-05	-6.35E-06	-3.91E-06	-5.74E-06	-4.57E-06	4.36E-06	5.75E-07	-2.05E-06	-1.42E-06	-4.44E-06	-1.37E-06
X_7	-1.52E-06	-5.77E-06	-3.4E-06	3.57E-06	1.44E-06	-2.98E-05	4.19E-05	3.61E-06	3.31E-06	-2.26E-06	-1.75E-06	-3.16E-06	-2.56E-06	1.4E-06	1.94E-06	-1.18E-07	-2.01E-06
X_8	-2.53E-06	4.98E-06	-8.65E-07	-1.88E-06	-2.54E-06	-6.35E-06	3.61E-06	1.76E-05	-2.07E-06	-7.31E-07	-2.35E-06	-1.35E-06	8.37E-07	-3.73E-07	2.92E-07	-1.87E-06	1.61E-07
X_9	-4E-06	-7.57E-06	-1.62E-06	1.16E-06	2.63E-06	-3.91E-06	3.31E-06	-2.07E-06	1.64E-05	1.32E-06	-2.86E-06	-1.56E-06	-4.48E-07	-8.37E-08	3.15E-06	2.06E-06	1.72E-06
X_{10}	-1.49E-06	-1.95E-05	-1.51E-06	9.35E-07	-4.5E-07	-5.74E-06	-2.26E-06	-7.31E-07	1.32E-06	4.36E-05	-1.64E-05	1.22E-05	-4.41E-06	-1.32E-05	5.07E-06	5.42E-07	-4.59E-06
X_{11}	3.26E-06	9.83E-06	9.22E-07	-2.06E-06	-1.6E-06	-4.57E-06	-1.75E-06	-2.35E-06	-2.86E-06	-1.64E-05	3.02E-05	-1.16E-05	-1.73E-08	9.91E-06	-8.37E-06	1.98E-06	-4.68E-06
X_{12}	4.43E-06	-2.53E-06	2.93E-06	3.41E-07	2.77E-06	4.36E-06	-3.16E-06	-1.35E-06	-1.56E-06	1.22E-05	-1.16E-05	0.000213	-3.85E-06	-0.000205	7.74E-06	3.6E-06	-4.77E-06
X_{13}	-3.73E-06	-1.01E-06	-1.27E-06	-6.46E-07	-3.09E-06	5.75E-07	-2.56E-06	8.37E-07	-4.48E-07	-4.41E-06	-1.73E-08	-3.85E-06	2.02E-05	-1.27E-06	5.62E-07	2.19E-06	-1.9E-06
X_{14}	-7.25E-06	1.78E-05	-3.99E-06	7.2E-07	-5.78E-06	-2.05E-06	1.4E-06	-3.73E-07	-8.37E-08	-1.32E-05	9.91E-06	-0.000205	-1.27E-06	0.000217	-1.09E-05	-6.08E-06	-2.41E-07
X_{15}	1.92E-06	-5.62E-05	-4.13E-06	4.17E-06	1.02E-06	-1.42E-06	1.94E-06	2.92E-07	3.15E-06	5.07E-06	-8.37E-06	7.74E-06	5.62E-07	-1.09E-05	7.12E-05	-4.34E-06	-9.08E-06
X_{16}	-3.63E-07	-4.16E-06	2.55E-06	-1.36E-06	1.42E-06	-4.44E-06	-1.18E-07	-1.87E-06	2.06E-06	5.42E-07	1.98E-06	3.6E-06	2.19E-06	-6.08E-06	-4.34E-06	1.64E-05	1.58E-06
X_{17}	2.6E-06	-2E-05	-3.04E-07	-3.89E-06	-5.64E-06	-1.37E-06	-2.01E-06	1.61E-07	1.72E-06	-4.59E-06	-4.68E-06	-4.77E-06	-1.9E-06	-2.41E-07	-9.08E-06	1.58E-06	4.73E-05

TABLE 9.4 MDs and MDAs for the Normal Group

Sample No.	1	2	3	4	5	6	7	8	...	196	197	198	199	200	Average
MD inverse	0.378374	0.431373	0.403562	0.500211	0.515396	0.495501	0.583142	0.565654	...	1.74	1.75	1.78	1.76	2.36	0.995
MD adjoint	0.000005	0.000006	0.000005	0.000007	0.000007	0.000007	0.000008	0.000007	...	0.00002	0.00002	0.00002	0.00002	0.00003	0.000013

TABLE 9.5 MDs and MDAs for Abnormals

Sample No.	1	2	3	4	5	6	7	8	...	13	14	15	16	17	Average
MD inverse	7.72741	8.41629	10.29148	7.20516	10.59075	10.55711	13.31775	14.81278	...	19.65543	43.04050	78.64045	97.27242	135.70578	30.39451
MD adjoint	0.00010	0.00011	0.00014	0.00009	0.00014	0.00014	0.00017	0.00019	...	0.00026	0.00057	0.00103	0.00128	0.00178	0.00040

TABLE 9.6   Dynamic *S/N* Ratios for the Combination of $L_{32}(2^{31})$ Array

Run	*S/N* Ratio	
	Inverse	Adjoint
1	−6.252	42.560
2	−6.119	42.693
3	−10.024	38.788
4	−10.181	38.631
5	−10.348	38.464
6	−10.495	38.317
7	−7.934	40.878
8	−8.177	40.635
9	−9.234	39.578
10	−9.631	39.181
11	−3.338	45.474
12	−3.406	45.406
13	−10.932	37.880
14	−11.121	37.691
15	−6.495	42.317
16	−7.265	41.547
17	−7.898	40.914
18	−7.665	41.147
19	−10.156	38.656
20	−9.901	38.911
21	−5.431	43.381
22	−5.312	43.500
23	−7.603	41.209
24	−7.498	41.314
25	−11.412	37.400
26	−11.100	37.712
27	−5.874	42.938
28	−4.989	43.823
29	−9.238	39.574
30	−8.989	39.823
31	−5.544	43.268
32	−5.303	43.509

TABLE 9.7   Gain in *S/N* Ratios (dB)

	Inverse Method				Adjoint Method		
Variable	Level 1	Level 2	Gain	Variable	Level 1	Level 2	Gain
X_1	−8.185	−7.745	−0.440	X_1	40.627	41.067	−0.440
X_2	−8.187	−7.742	−0.445	X_2	40.625	41.070	−0.445
X_3	−8.249	−7.680	−0.569	X_3	40.563	41.132	−0.569
X_4	−7.949	−7.980	0.031	X_4	40.863	40.832	0.031
X_5	−7.069	−8.860	1.791	X_5	41.743	39.952	1.791
X_6	−8.318	−7.611	−0.706	X_6	40.494	41.201	−0.706
X_7	−7.976	−7.954	−0.022	X_7	40.836	40.858	−0.022
X_8	−8.824	−7.105	−1.718	X_8	39.988	41.707	−1.718
X_9	−8.188	−7.742	−0.446	X_9	40.625	41.070	−0.446
X_{10}	−6.358	−9.571	3.212	X_{10}	42.454	39.241	3.212
X_{11}	−8.101	−7.828	−0.273	X_{11}	40.711	40.984	−0.273
X_{12}	−7.821	−8.108	0.287	X_{12}	40.991	40.704	0.287
X_{13}	−7.562	−8.367	0.805	X_{13}	41.250	40.445	0.805
X_{14}	−7.315	−8.615	1.300	X_{14}	41.497	40.197	1.300
X_{15}	−7.590	−8.339	0.749	X_{15}	41.222	40.473	0.749
X_{16}	−7.982	−7.947	−0.035	X_{16}	40.830	40.865	−0.035
X_{17}	−7.832	−8.097	0.265	X_{17}	40.980	40.715	0.265

9.2.2   Example 2

This case study was related to the *process of diagnosis of babies after delivery* and was carried out under an American Society for Quality (ASQ) research fellowship grant program for the year 1999. This process was considered to be critical, as there were several instances of false alarms. The existing diagnosis process was based only on a weight consideration; other variables were not considered. According to this practice, babies weighing between 2500 and 4000 g are considered normal or healthy babies as shown in Table 9.8. After a series of discussions, it was decided to use 12 variables (X_1, X_2, ..., X_{12}) for constructing the measurement scale.

TABLE 9.8   Existing Practice

Weight (g)	Condition
<2500	Unhealthy
Between 2500 and 4000	Healthy
>4000	Unhealthy

For constructing the Mahalanobis space, babies weighing 3000 g $\pm$ 5% are considered. In this example, there are 58 normals and 30 abnormals. The MDs corresponding to these babies are computed using MTS/MTGS methods. The MTGS method gave better results than the MTS method: the average MD with the MTGS method is 0.98, and with the MTS method it is 0.92. The reason for this discrepancy is the existence of multicollinearity. This is clear from the correlation matrix (Table 9.9), which shows that the variables X_{10}, X_{11}, and X_{12} have high correlations with each other. The determinant of the matrix is also estimated and found to be 8.693×10^{-12} (close to zero), indicating that the matrix is almost singular. The presence of multicollinearity will also affect the other stages of the MTS method. Hence, MTGS method and adjoint matrix method are used separately to perform the rest of the analysis.

MTGS Method

To ensure the goodness of the measurement scale, MDs of 30 abnormals are computed. The babies weighing less than 2850 g and more than 3150 g are considered abnormals. The MDs of the abnormals are high enough to ensure the accuracy of the measurement scale.

Identification of Useful Variables

The S/N ratios of the variables are computed based on larger-the-better-type criterion. After conducting S/N ratio analysis, useful variables are identified. Table 9.10 gives S/N ratios associated with the variables. From this table, the combination X_1-X_3-X_4-X_6-X_7-X_{10}-X_{11}-X_{12} is considered to be important, because these variables have higher S/N ratios than others. With the useful variables, a confirmation run is conducted. The results (Figure 9.1) show that abnormal MDs have higher values and the measurement scale with all the variables is able to make a clear distinction between normals and abnormals.

Adjoint Matrix Method

For this example, the adjoint matrix method is applied. The adjoint of the correlation matrix is shown in Table 9.11. After computing MDAs for normals, the measurement scale is validated by com-

TABLE 9.9 Correlation Matrix

	X_1	X_2	X_3	X_4	X_5	X_6	X_7	X_8	X_9	X_{10}	X_{11}	X_{12}
X_1	1	0.358	-0.085	-0.024	0.005	0.057	-0.149	-0.128	-0.046	0.105	-0.055	-0.055
X_2	0.358	1	0.014	-0.022	0.003	-0.097	-0.271	-0.079	0.061	0.325	0.023	0.023
X_3	-0.085	0.014	1	0.0769	0.0708	0.0577	0.3138	0.1603	0.0815	0.4945	0.5286	0.5333
X_4	-0.024	-0.022	0.0769	1	-0.135	-0.018	0.296	-0.206	0.062	0.597	0.624	0.622
X_5	0.005	0.003	0.0708	-0.135	1	0.123	0.264	0.114	0.053	0.536	0.560	0.559
X_6	0.057	-0.097	0.0577	-0.018	0.123	1	0.353	0.055	0.056	0.063	0.096	0.096
X_7	-0.149	-0.271	0.3138	0.296	0.264	0.353	1	0.103	0.092	0.395	0.508	0.508
X_8	-0.128	-0.079	0.1603	-0.206	0.114	0.055	0.103	1	-0.153	-0.032	-0.002	-0.0004
X_9	-0.046	0.061	0.0815	0.062	0.053	0.056	0.092	-0.153	1	0.116	0.104	0.104
X_{10}	0.105	0.325	0.4945	0.597	0.536	0.063	0.395	-0.032	0.116	1	0.951	0.951
X_{11}	-0.055	0.023	0.5286	0.624	0.560	0.096	0.508	-0.002	0.104	0.951	1	0.999
X_{12}	-0.055	0.023	0.5333	0.622	0.559	0.096	0.508	-0.0004	0.104	0.951	0.999	1

TABLE 9.10   *S/N* **Ratios (dB) of the Variables**

X_1	X_2	X_3	X_4	X_5	X_6	X_7	X_8	X_9	X_{10}	X_{11}	X_{12}
−0.03	−39.11	2.91	−4.11	−40.52	−5.11	−4.22	−15.58	−40.09	0.12	3.62	2.48

puting abnormal MDAs. Figure 9.2 indicates that there is a clear distinction between normals and abnormals.

In the next step, important variables are selected using the $L_{16}(2^{15})$ array. The *S/N* ratio analysis was performed based on larger-the-better criterion. The gains in *S/N* ratios are shown in Table 9.12. From this table, it is clear that the variables X_1-X_2-X_3-X_4-X_6-X_{10}-X_{11}-X_{12} have positive gains and, hence, they are important. The confirmation run with these variables (Figure 9.3) indicates that distinction (between normals and abnormals) is better than that with the initial condition having all variables. Therefore, the adjoint matrix method can safely replace the inverse matrix method, since it is as efficient as the inverse matrix method, in general, and more efficient when there are problems of multi-collinearity. However, MTGS method is required to identify the direction of abnormals.

9.3   β-ADJUSTMENT METHOD FOR SMALL CORRELATIONS

So far, we have discussed about the treatment of datasets with multicollinearity or high correlations. Another important situation to be considered in multivariate analysis is the presence of small

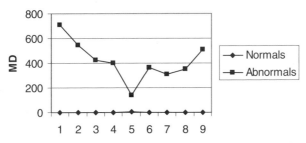

Figure 9.1   MDs for normals versus abnormals with useful variables (only nine points are shown).

TABLE 9.11 Adjoint Matrix

	X_1	X_2	X_3	X_4	X_5	X_6	X_7	X_8	X_9	X_{10}	X_{11}	X_{12}
X_1	1.00912E-10	4.70272E-10	1.61623E-10	2.76032E-10	2.57713E-10	-5.48951E-12	5.043E-12	7.14809E-13	1.43647E-12	-1.66567E-09	7.66095E-10	4.08691E-10
X_2	4.70263E-10	2.50034E-09	9.18237E-10	1.55621E-09	1.45406E-09	-2.10511E-11	2.83118E-11	-3.09613E-12	8.03373E-13	-8.7444E-09	4.41674E-09	1.58527E-09
X_3	1.61527E-10	9.17746E-10	1.06463E-09	1.63137E-09	1.50922E-09	5.28862E-13	2.04944E-11	-9.18812E-12	-1.3292E-11	-3.18575E-09	5.68418E-09	-5.10159E-09
X_4	2.7594E-10	1.55576E-09	1.63154E-09	2.56985E-09	2.37158E-09	-3.57245E-13	3.50392E-11	-1.10855E-11	-1.89581E-11	-5.40857E-09	6.24469E-09	-4.93127E-09
Z_5	2.57631E-10	1.45366E-09	1.50939E-09	2.37159E-09	2.20389E-09	-1.73783E-12	3.34823E-11	-1.29848E-11	-1.78615E-11	-5.0537E-09	5.64554E-09	-4.38529E-09
X_6	-5.4903E-12	-2.10556E-11	5.23064E-13	-3.64155E-13	-1.74411E-12	1.06058E-11	-4.37752E-12	-1.97695E-13	-5.79622E-13	7.5335E-11	3.17881E-13	-6.9595E-11
X_7	5.04604E-12	2.83284E-11	2.05079E-11	3.50574E-11	3.34989E-11	-4.37759E-12	1.58563E-11	-1.42556E-12	-1.00253E-12	-8.62928E-11	-1.25906E-10	1.486E-10
X_8	7.12086E-13	-3.11071E-12	-9.19606E-12	-1.10978E-11	-1.29962E-11	-1.97598E-13	-1.42569E-12	1.01743E-11	1.84668E-11	1.04492E-11	1.34899E-10	-1.25096E-10
X_9	1.43722E-12	8.07304E-13	-1.3290E-11	-1.89556E-11	-1.78591E-11	-5.79657E-13	-1.00246E-12	1.84666E-12	9.46854E-12	-6.93471E-12	-2.47767E-11	5.98708E-11
X_{10}	-1.66565E-09	-8.74446E-09	-3.1875E-09	-5.4102E-09	-5.05514E-09	7.53194E-11	-8.62349E-11	1.03982E-11	-6.92086E-12	3.07209E-08	-1.50768E-08	-6.10343E-09
X_{11}	7.60305E-10	4.38609E-09	5.67096E-09	6.22205E-09	5.62443E-09	5.56545E-13	-1.26294E-10	1.35001E-10	-2.47494E-11	-1.49692E-08	2.88114E-07	-2.83899E-07
X_{12}	4.14615E-10	1.61673E-09	5.08692E-09	-4.90701E-09	-4.36272E-09	-6.98298E-11	1.48962E-10	-1.25168E-10	5.98339E-11	-6.21375E-11	-2.8383E-07	2.97854E-07

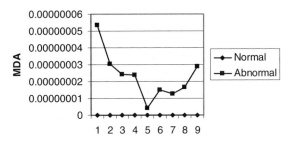

Figure 9.2 MDAs for normals versus abnormals (only nine points are shown).

correlation coefficients in correlation matrix. When there are small correlation coefficients, the adjustment factor β is calculated as follows:

$$
\beta = \begin{cases} 0 & \text{if } |r| \leq \dfrac{1}{\sqrt{n}} \\[2ex] 1 - \left(\dfrac{1}{n-1}\right)\left(\dfrac{1}{r^2} - 1\right) & \text{if } |r| > \dfrac{1}{\sqrt{n}} \end{cases} \tag{9.4}
$$

where r is correlation coefficient, n is sample size, and $|r|$ is absolute value of r. After β is computed, the elements of the correlation matrix are adjusted by multiplying them with β. This adjusted matrix is used to carry out MTS analysis or analysis with the adjoint matrix.

TABLE 9.12 Gain in S/N Ratio (dB)

Variable	Level 1	Level 2	Gain
X_1	−102.90	−105.01	2.12
X_2	−103.53	−104.38	0.86
X_3	−103.84	−104.07	0.22
X_4	−103.72	−104.19	0.47
X_5	−104.04	−103.86	−0.18
X_6	−103.87	−104.04	0.16
X_7	−104.18	−103.72	−0.46
X_8	−104.14	−103.77	−0.37
X_9	−104.33	−103.58	−0.76
X_{10}	−103.51	−104.40	0.90
X_{11}	−103.78	−104.13	0.35
X_{12}	−103.43	−104.48	1.05

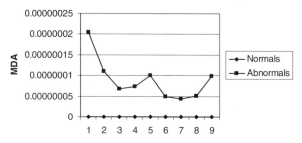

Figure 9.3   MDAs for normals versus abnormals with useful variables (only nine points are shown).

The β-adjustment method is applied to the medical diagnosis example (with 17 variables) since the correlation matrix in this case contains a few small correlation coefficients. The β-adjusted correlation matrix (using Equation 9.4) is as shown in Table 9.13. With this matrix, MTS analysis is carried out and as a result the following useful variable combination is obtained: X_2-X_5-X_6-X_7-X_{10}-X_{12}-X_{13}-X_{14}-X_{15}. Here S/N ratios are calculated based on larger-the-better criterion. This combination is very similar to the useful variable set obtained without β-adjustment (Chapter 2); the only difference is presence of variable X_6. With this useful variable set, S/N ratio analysis is carried out to measure the improvement in overall system performance. From Table 9.14, which shows system performance in S/N ratios, it is clear that there is a gain of 2.07 dB if useful variables are used instead of entire set of variables.

9.4   SUBSET SELECTION USING THE MULTIPLE MAHALANOBIS DISTANCE METHOD

Selection of suitable subsets is very important in multivariate diagnosis/pattern recognition activities, because it is difficult to handle large datasets with several number of variables. We know that in MTS/MTGS methods subset selection is done using S/N ratios obtained from the conditions outside MS. In this section a new metric, called multiple Mahalanobis distance (MMD), is introduced to compute S/N ratios and thereby select suitable subsets. This method is very useful in complex situations, where we

TABLE 9.13 β-Adjusted Correlation Matrix

	X_1	X_2	X_3	X_4	X_5	X_6	X_7	X_8	X_9	X_{10}	X_{11}	X_{12}	X_{13}	X_{14}	X_{15}	X_{16}	X_{17}
X_1	1.000	-0.281	-0.261	-0.392	-0.199	0.052	0.000	0.185	0.277	-0.056	-0.067	0.247	0.099	0.267	-0.276	0.000	-0.265
X_2	-0.281	1.000	0.055	0.406	0.687	0.271	0.368	-0.061	0.000	0.643	0.384	-0.217	0.252	-0.201	0.885	0.236	0.796
X_3	-0.261	0.055	1.000	0.417	0.178	0.024	0.103	0.002	0.000	0.149	0.155	0.000	0.127	0.014	0.117	-0.078	0.173
X_4	-0.392	0.406	0.417	1.000	0.301	0.000	0.000	0.000	-0.059	0.252	0.197	-0.100	0.050	-0.099	0.353	0.036	0.403
X_5	-0.199	0.687	0.178	0.301	1.000	0.332	0.374	0.000	0.000	0.572	0.419	0.000	0.355	0.000	0.640	0.099	0.671
X_6	0.052	0.271	0.024	0.000	0.332	1.000	0.788	0.301	0.149	0.544	0.528	0.115	0.305	0.139	0.307	0.154	0.347
X_7	0.000	0.368	0.103	0.000	0.374	0.788	1.000	0.109	0.000	0.562	0.500	0.097	0.362	0.115	0.387	0.064	0.425
X_8	0.185	-0.061	0.002	0.000	0.000	0.301	0.109	1.000	0.208	0.090	0.206	0.231	0.054	0.238	0.000	0.043	0.000
X_9	0.277	0.000	0.000	-0.059	0.000	0.149	0.000	0.208	1.000	0.000	0.113	0.143	0.080	0.139	-0.007	-0.044	0.000
X_{10}	-0.056	0.643	0.149	0.252	0.572	0.544	0.562	0.090	0.000	1.000	0.679	0.000	0.427	0.016	0.607	0.103	0.645
X_{11}	-0.067	0.384	0.155	0.197	0.419	0.528	0.500	0.206	0.113	0.679	1.000	0.128	0.329	0.120	0.436	0.000	0.457
X_{12}	0.247	-0.217	0.000	-0.100	0.000	0.115	0.097	0.231	0.143	0.000	0.128	1.000	0.296	0.966	-0.105	0.000	0.000
X_{13}	0.099	0.252	0.127	0.050	0.355	0.305	0.362	0.054	0.080	0.427	0.329	0.296	1.000	0.304	0.249	0.000	0.339
X_{14}	0.267	-0.201	0.014	-0.099	0.000	0.139	0.115	0.238	0.139	0.016	0.120	0.966	0.304	1.000	-0.077	0.000	0.000
X_{15}	-0.276	0.885	0.117	0.353	0.640	0.307	0.387	0.000	-0.007	0.607	0.436	-0.105	0.249	-0.077	1.000	0.262	0.768
X_{16}	0.000	0.236	-0.078	0.036	0.099	0.154	0.064	0.043	-0.044	0.103	0.000	0.000	0.000	0.000	0.262	1.000	0.149
X_{17}	-0.265	0.796	0.173	0.403	0.671	0.347	0.425	0.000	0.000	0.645	0.457	0.000	0.339	0.000	0.768	0.149	1.000

TABLE 9.14   *S/N* **Ratio Analysis (β-Adjustment Method)**

S/N ratio, optimal system	13.40 dB
S/N ratio, original system	11.33 dB
Gain	2.07 dB

have problems like voice recognition or face recognition. In these cases, the number of variables runs into the order of several hundreds. Because of this, it is difficult to conduct MTS/MTGS analysis on all variables to find variables of importance. Use of the MMD method will reduce problem complexity and help one make effective decisions.

In the MMD method, the large number of variables is divided into several subsets containing local variables. For example, in a voice recognition system (as shown in Figure 9.4), let there be k subsets. The subsets correspond to k patterns numbered from 1,2,..k. Each pattern starts at a low value, reaches a maximum, and then returns to the low value. These patterns (subsets) are described by a set of respective local variables. In the MMD method, the Mahalanobis distances are calculated for each subset. The Mahalanobis distances are used to calculate the MMD. Using abnormal MMDs, *S/N* ratios are calculated to determine useful subsets. In this way the complexity of the problems is reduced.

This method is useful for identifying the subsets (or variables in the subsets) corresponding to different failure modes or patterns that are responsible for higher values of MDs. For example, in the case of a final product inspection system, use of the MMD method would help to find variables corresponding to different processes that are responsible for product failure. If the variables corresponding to different subsets or processes cannot be identified,

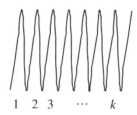

Figure 9.4   Example of voice recognition patterns.

then the decision maker can select subsets from the original set of variables and identify the most appropriate subsets.

9.4.1 Steps in the MMD Method

1. Define subsets from original set of variables. The subsets may contain variables corresponding to different patterns or failure modes. These variables can also be selected based on the decision maker's discretion. The number of variables in the subsets need not be the same.
2. For each subset, calculate MDs (for normals and abnormals) using respective variables in them.
3. Compute square root of these MDs ($\sqrt{\text{MDs}}$).
4. Consider the subsets as variables (control factors). The $\sqrt{\text{MDs}}$ would provide required data for these subsets. If there are k subsets, then the problem is similar to a multivariate problem with k variables. The number of normals and abnormals will be same as in the original problem. The analysis with $\sqrt{\text{MDs}}$ is exactly the same as that of the MTS method or adjoint matrix method with original variables. The new Mahalanobis distance obtained based on square root of MDs is referred to as multiple Mahalanobis distance.
5. With the MMDs, S/N ratios are obtained for each run of an orthogonal array. Based on gains in S/N ratios, the important subsets are selected.

9.4.2 Example

MMD analysis is performed for the medical diagnosis example (with 17 variables) illustrated in earlier chapters. From the 17 variables, eight subsets (Table 9.15) are selected to illustrate the methodology; there is no rationale for this selection. Note that the number of variables in each subset is not the same.

For each subset, Mahalanobis distances are computed with the help of correlation matrices corresponding to the respective variables. Therefore, we have eight sets of MDs (for normals and abnormals) corresponding to the subsets. The $\sqrt{\text{MDs}}$ provide data

TABLE 9.15   Subsets for MMD Analysis

Subset	Variables
S_1	X_1-X_2-X_3-X_4
S_2	X_5-X_6-X_7-X_8
S_3	X_9-X_{10}-X_{11}-X_{12}
S_4	X_{13}-X_{14}-X_{15}-X_{16}-X_{17}
S_5	X_3-X_4-X_5-X_6
S_6	X_{10}-X_{11}-X_{12}-X_{13}-X_{14}-X_{15}
S_7	X_{14}-X_{15}-X_{16}-X_{17}
S_8	X_2-X_5-X_7-X_{10}-X_{12}-X_{13}-X_{14}-X_{15}

corresponding to the subsets for MMD analysis. Tables 9.16 and 9.17 show sample data ($\sqrt{\text{MD}}$s) for MMD analysis.

After arranging the data ($\sqrt{\text{MD}}$s) in this manner, MMD analysis is carried out. In this analysis, MMDs are Mahalanobis distances (scaled distances) obtained from $\sqrt{\text{MD}}$s. Tables 9.18 and 9.19 provide sample values of MMDs for normals and abnormals respectively.

The next step is to assign the subsets to the columns of a suitable orthogonal array. Since there are eight subsets, the $L_{12}(2^{11})$ array was selected. The abnormal MMDs are computed for each run of this array. After performing average response analysis, gains in larger-the-better-type S/N ratios are computed for all the subsets. These details are shown in Table 9.20. From this table it

TABLE 9.16 $\sqrt{\text{MD}}$s for Normals (Sample Data)

Sample No.	S_1	S_2	S_3	S_4	S_5	S_6	S_7	S_8
1	0.873	0.545	0.707	0.756	0.796	0.505	0.832	0.574
2	0.762	0.540	0.929	0.710	0.499	0.688	0.606	0.807
3	1.022	0.688	0.550	0.623	0.955	0.479	0.697	0.613
4	1.102	0.544	0.769	0.740	1.225	0.648	0.827	0.681
5	1.022	0.640	0.602	0.888	0.815	0.782	0.934	0.695
⋮								
196	1.041	0.786	1.691	1.513	0.500	1.550	1.539	1.411
197	1.467	1.310	2.101	1.201	1.457	1.481	0.611	1.373
198	1.086	1.278	0.974	1.406	1.410	1.834	0.994	1.648
199	1.238	0.999	1.107	1.061	1.206	1.132	0.964	1.700
200	1.391	0.924	0.979	0.680	1.094	2.156	0.750	1.844

TABLE 9.17 $\sqrt{\text{MDs}}$ for Abnormals (Sample Data)

Sample No.	S_1	S_2	S_3	S_4	S_5	S_6	S_7	S_8
1	1.339	2.930	2.610	3.428	2.574	3.277	2.913	3.734
2	1.491	3.469	1.931	1.511	3.267	3.388	1.687	3.932
3	1.251	2.700	0.742	2.631	2.447	3.322	2.660	4.365
4	2.124	2.507	2.041	3.240	2.518	3.058	2.009	3.395
5	1.010	2.182	2.867	1.279	1.861	4.035	1.090	4.440
$\vdots$								
13	1.769	2.819	6.544	2.153	2.352	6.023	2.177	5.776
14	1.898	2.045	3.817	4.551	2.443	10.213	1.969	9.275
15	1.624	12.681	2.116	3.672	12.248	9.064	1.202	11.426
16	5.453	13.314	3.630	1.022	13.515	10.095	1.108	12.121
17	4.511	16.425	5.489	3.684	12.027	11.142	2.264	10.939

is clear that S_8 has the highest gain, indicating that this is an important subset. The variables in this subset are same as the useful variables obtained from the MTS method (please see Chapter 2 or Chapter 4). Ideally, the subsets (or the variables in the subsets) with positive gains must be considered useful for diagnosis. This example is a simple case where we have only 17 variables and, therefore, here, the MMD method may not be necessary. However, in complex cases, with several hundreds of variables, the MMD method is more appropriate and reliable.

Note. We can also use the Gram–Schmidt method or adjoint matrix method to calculate MMDs. With the Gram–Schmidt method, Gram–Schmidt coefficients are used instead of the correlation matrix, and with the adjoint matrix method, the adjoint matrix is used instead of the correlation matrix.

9.5   SELECTION OF MAHALANOBIS SPACE FROM HISTORICAL DATA

We know that selection of a good Mahalanobis space (MS) is important for the MTS, MTGS, and adjoint matrix methods. Although the decision makers can use their discretion to make a judgment about MS, sometimes it may be hard to select MS, especially when we are dealing with historical data, by which

TABLE 9.18 MMDs for Normals (Sample Values)

Condition	1	2	3	4	5	6	7	8	9	10	...	198	199	200
MMD	0.558	0.861	0.425	0.786	0.413	1.655	0.357	0.660	0.641	0.717	...	2.243	2.243	4.979

TABLE 9.19 MMDs for Abnormals (Sample Values)

Condition	1	2	3	4	5	6	7	8	9	10	...	15	16	17
MMD	22.52	29.86	30.61	23.47	27.05	57.12	61.61	52.64	50.77	66.15	...	515.50	601.30	592.37

TABLE 9.20   Gain in *S/N* Ratios (dB)

	Level 1	Level 2	Gain
S_1	15.498	18.053	−2.555
S_2	17.463	16.089	1.374
S_3	16.712	16.839	−0.127
S_4	15.925	17.627	−1.702
S_5	17.626	15.926	1.700
S_6	17.243	16.309	0.934
S_7	15.683	17.869	−2.186
S_8	18.556	14.996	3.560

"healthiness" of a condition cannot be defined. In such situations, to start with, we can generate the MS and calculate the MDs of all observations in the data. Then we discard the observations having higher MDs and recalculate MDs with a new MS constructed with the remaining observations. This process is repeated until we generate a suitable MS for conducting multivariate diagnosis. While carrying out this process, it is important to have an adequate sample size. Here, also, the role of the decision maker is significant in identifying a good MS through this iterative process.

9.6   CONCLUSIONS

- This chapter introduces the advanced topics of MTS/MTGS methods.
- The adjoint matrix method is an effective way of conducting a diagnosis process. This method works as well as the MTS method, in general, and is more efficient than MTS in the presence of multicollinearity. However, the MTGS method has to be used to find the direction of abnormals.
- The multiple Mahalanobis distance method is a flexible and easy technique that can be used to reduce problem complexity and make effective decisions.
- The β-adjustment method is used in situations where there are small correlation coefficients. The β-adjusted correlation matrix is used to perform MTS or the adjoint matrix method.
- The iterative process described in Section 9.5 is helpful in selecting a suitable Mahalanobis space with historical data by which "healthiness" of a condition cannot be defined.

10
MTS/MTGS VERSUS OTHER METHODS

This chapter is important because it clarifies and explains the differences between the other methods of multivariate analysis or pattern recognition and MTS/MTGS. For the purpose of this comparison, the medical diagnosis data (liver disease test data) discussed in previous chapters are used. MTS/MTGS methods are compared with the following multivariate techniques:

- Principal component analysis
- Discrimination and classification method
- Multiple regression analysis
- Stepwise regression
- Test of additional information (Rao's test)
- Multivariate control charts
- Artificial neural networks

Basis for Comparison

Unlike many multivariate methods, MTS/MTGS methods are data analytic and use simple measures of descriptive statistics; they are not probabilistic in nature. Artificial Neural Networks (ANN) are also data analytic techniques. MTS/MTGS methods are different from other methods (including ANN) in the following ways:

- Mahalanobis distance* is suitably scaled and used for measuring the degree of abnormality on a measurement scale. The base or reference point of the scale is obtained from the Mahalanobis space (MS). Mahalanobis space, which depends upon decision maker's discretion, can be constructed for all types of variables irrespective of their distributions.
- Each abnormal condition (or the condition outsite MS) is considered unique and abnormals are not considered a separate population.

The list of variables corresponding to the liver disease test data from Chapter 2 is repeated in Table 10.1. The software programs MINITAB and Microsoft Excel are used to perform the multivariate analysis using classical methods, and MATLAB is used for conducting analysis with ANN.

TABLE 10.1 Variables in Medical Diagnosis Data

Sample No.	Variables	Notation	Notation for Analysis
1	Age		X_1
2	Sex		X_2
3	Total protein in blood	TP	X_3
4	Albumin in blood	Alb	X_4
5	Cholinesterase	ChE	X_5
6	Glutamate O transaminase	GOT	X_6
7	Glutamate P transaminase	GPT	X_7
8	Lactate dehydrogenase	LHD	X_8
9	Alkanline phosphatase	Alp	X_9
10	r-Glutamyl transpeptidase	r-GPT	X_{10}
11	Leucine aminopeptidase	LAP	X_{11}
12	Total cholesterol	TCh	X_{12}
13	Triglyceride	TG	X_{13}
14	Phospholopid	PL	X_{14}
15	Creatinine	Cr	X_{15}
16	Blood urea nitrogen	BUN	X_{16}
17	Uric acid	UA	X_{17}

*Since Mahalanobis distance (in its original form) is used in classical methods, in this chapter, it is denoted as MD. The scaled Mahalanobis distance is denoted as scaled MD.

10.1 PRINCIPAL COMPONENT ANALYSIS

Principal component analysis (PCA) aims at explaining the variance–covariance structure through fewer linear combinations of original variables. Johnson and Wichern (1992) provide a clear discussion of PCA. These linear combinations are called the principal components (PCs). The principal components depend entirely on the covariance or correlation matrix. Their development does not require multivariate normality assumption.

Let the standardized random vector $Z^T = [Z_1, Z_2, ..., Z_p]$ have the correlation matrix C with eigenvalues $\lambda_1 \geq \lambda_2 \geq \cdots \geq \lambda_p \geq 0$, where T denotes transpose of the vector. Consider the following linear combinations:

$$Y_1 = l_1^T Z = l_{11}Z_1 + l_{21}Z_2 + \cdots + l_{p1}Z_p$$

$$Y_2 = l_2^T Z = l_{12}Z_1 + l_{22}Z_2 + \cdots + l_{p2}Z_p$$

$$\vdots$$

$$Y_p = l_p^T Z = l_1 p Z_1 + l_{2p}Z_2 + \cdots + l_{pp}Z_p$$

We can write

$$V(Y_i) = l_i^T C \, l_i \quad \text{and} \quad \text{Cov}(Y_i \, Y_k) = l_i^T C \, l_k$$

where $i,k = 1, 2, ..., p$.

The principal components are those uncorrelated linear combinations $Y_1, Y_2, ..., Y_p$ whose variances $[V(Y_i)\text{s}]$ are as large as possible.

We can also prove that the total variance $= \Sigma_{i=1}^p V(Z_i) = \lambda_1 + \lambda_2 + \cdots + \lambda_p = \Sigma_{i=1}^p V(Y_i)$. The proportion of total variance due to the kth principal component $= \lambda_k/(\lambda_1 + \lambda_2 + \cdots + \lambda_p)$. Note that the principal components can be constructed either by covariance matrix or correlation matrix, depending on the measurement units of variables. Since the variables in medical diagnosis data have different units, a correlation matrix based on standardized values of all observations (normals and abnormals) is used for performing PCA.

The eigen analysis of the principal components is given in Table 10.2. The linear combinations corresponding to the fifteen prin-

TABLE 10.2   Eigen Analysis of Principle Components (PCs)

PC	Y_1	Y_2	Y_3	Y_4	Y_5	Y_6
Eigenvalue	4.8479	3.6445	2.1888	1.1344	0.9283	0.7909
Proportion	0.285	0.214	0.129	0.067	0.055	0.047
Cumulative	0.285	0.500	0.628	0.695	0.750	0.796

PC	Y_7	Y_8	Y_9	Y_{10}	Y_{11}	Y_{12}
Eigenvalue	0.6256	0.5414	0.4968	0.4673	0.3469	0.3247
Proportion	0.037	0.032	0.029	0.027	0.020	0.019
Cumulative	0.833	0.865	0.894	0.922	0.942	0.961

PC	Y_{13}	Y_{14}	Y_{15}	Y_{16}	Y_{17}
Eigenvalue	0.2483	0.1987	0.1063	0.0705	0.0386
Proportion	0.015	0.012	0.006	0.004	0.002
Cumulative	0.976	0.987	0.994	0.998	1.000

cipal components are given in Appendix 2. The first component has the form

$$Y_1 = -0.166Z_1 - 0.131Z_2 + 0.169Z_3 + 0.222Z_4 + 0.217Z_5$$
$$- 0.364Z_6 - 0.359Z_7 - 0.317Z_8 - 0.191Z_9 - 0.347Z_{10}$$
$$- 0.247Z_{11} - 0.100Z_{12} - 0.327Z_{13} - 0.282Z_{14}$$
$$- 0.151Z_{15} - 0.023Z_{16} - 0.174Z_{17}$$

Similarly, we can obtain other combinations. Using these principal components, transformations corresponding to original variables can be obtained. Usually, if $k < p$ components account for 90% of the total variability, then these components are sufficient to explain variability (correlation) structure. Since the eigenvalues associated with the correlation matrix represent the variability due to the variables, from Table 10.2 it is clear that first 10 variables account for 92.2% of total variability. Therefore, according to this analysis, the first 10 principal components are sufficient to describe the variability structure of the given system.

Usually, the method of conducting principal component analysis is carried out for dimensionality reduction, while explaining the variability of the system. Though fewer PCs are sufficient to explain variability structure, we still need all original variables to compute these PCs.

Since PCs are uncorrelated linear combinations of original variables, they are similar to orthogonal Gram–Schmidt vectors. Although, the primary objectives of MTS/MTGS methods are different from PCA, we can still compute MDs (or scaled MDs) using PCs in a similar way as in the MTGS method and follow the different stages of MTS/MTGS methods. We don't recommend this, because the basic objectives of MTS/MTGS are different and there is no need to calculate principal components, which involves additional computations.

The differences between MTS/MTGS and PCA are as follows:

- PCA reduces dimensionality in terms of linear combinations of original variables. To compute these linear combinations we need all original variables. PCA does not provide any means of reducing the dimensionality in terms of the original variables.

- Usually, selection of important principal components depends on the correlation matrix, which is based on a population. In MTS/MTGS methods, one of the primary objectives is to provide a measurement scale for multivariate systems. The Mahalanobis space (which contains means, SDs, and the correlation structure of the healthy group) provides a reference point to the scale. The dimensionality reduction is done based on ability of the scale to measure the conditions outside Mahalanobis space.

- In MTS/MTGS methods, use of OAs helps in the dimensionality reduction in terms of the original variables. Dimensionality reduction is based on MDs and S/N ratios. The MDs are obtained from the means, SDs, and the correlation matrix of the healthy group or Mahalanobis space.

10.2 DISCRIMINATION AND CLASSIFICATION METHOD

The discrimination and classification method is intended for separating the distinct sets of objects (or observations) and allocating new objects to previously defined groups. The objectives of the discrimination and classification method are as follows:

1. To describe the differential features of objects or observations from several known groups (populations). In this method "discriminants" are found in such a way that the groups are separated as much as possible.
2. To classify objects (observations) into two or more labeled classes. The emphasis is on deriving a rule that can be used to assign a new object to the labeled classes.

An extensive discussion on this method is provided in Johnson and Wichern (1992).

10.2.1 Fisher's Discriminant Function

Fisher's linear discriminant function is obtained by transforming the multivariate observations X to univariate observations Y such that the Ys derived from populations A and B are separated as much as possible.

Let there be two populations A and B. The linear discrimination function Y is calculated as

$$Y = L^T X,$$

where

$$L^T = (m_A - m_B) \, C^{-1}_{pooled}$$
$$m_A = \text{mean of } A$$
$$m_B = \text{mean of } B$$
$$T = \text{transpose of a vector}$$
$$X = \text{new observation}$$
$$C_{pooled} = \text{pooled covariance matrix}$$
$$C_{pooled} = \frac{1}{N_1 + N_2 - 2} [(N_1 - 1)C_1 + (N_2 - 1)C_2]$$
$$C_1 = \text{covariance matrix corresponding to the first group}$$
$$C_2 = \text{covariance matrix corresponding to the second group}$$
$$N_1 = \text{sample size of the first group}$$
$$N_2 = \text{sample size of the second group}$$

The cutoff point m is calculated as

$$m = \tfrac{1}{2}(m_A - m_B)C^{-1}{}_{\text{pooled}}(m_A + m_B)$$

If $L^T X > m$, classify X as population A

If $L^T X < m$, classify X as population B

10.2.2 Use of Mahalanobis Distance

An observation is classified into a group if the Mahalanobis distance (squared distance) of observation to the group center (mean) is the minimum. An assumption is made that covariance matrices are equal for all groups. There is a unique part of the squared distance formula for each group, which is called the linear discriminant function for that group. For any observation, the group with the smallest squared distance has the largest linear discriminant function and the observation is then classified into this group.

The maximum separation between the groups is given by

$$D^2 = (m_A - m_B)^T C^{-1}{}_{\text{pooled}}(m_A - m_B)$$

When there are more than two populations, we have the following decision rule:

- Calculate the Mahalanobis distance (D_i^2) associated with the observation X corresponding to a population i as $D_i^2 = (X - m_i)^T C^{-1}{}_{\text{pooled}}(X - m_i)$, where m_i = the mean of population i and C_{pooled} = the pooled covariance matrix.
- When there are k populations, assign X to group i ($i = 1, ..., k$), if $D_i^2 = \min(D_1^2, D_2^2, D_3^2, ..., D_k^2)$.

The Mahalanobis distance can also be written as

$$D_i^2 = -2[m_i^T C^{-1}{}_{\text{pooled}} X - 0.5 m_i^T C^{-1}{}_{\text{pooled}} m_i] + X^T C^{-1}{}_{\text{pooled}} X \tag{10.1}$$

where the term in square brackets is a linear function of X, and is called the linear discriminant function for group i. For a given

X, the group with the smallest squared distance has the largest linear discriminant function, which gives us another rule to classify an observation. According to this rule, assign X to a group having the largest discriminant function.

Discriminant analysis is performed for the medical diagnosis data based on covariance matrices. In this analysis, the abnormals are also considered a separate population. The sample size for the healthy group is 200 and for the abnormal group is 17. The linear discriminant functions for the two groups are given in Table 10.3.

TABLE 10.3   Linear Discriminant Functions and Misclassification Summary

Summary of Classification		
Put into	....True group....	
Group	1	2
1	200	0
2	0	17
Total N	200	17
N Correct	200	17
Proportion	1.000	1.000

$N = 217$ N correct $= 217$ Proportion correct $= 1.000$

Linear Discriminant Function for Groups

	1	2
Constant	-650.62	-679.79
X_1	1.42	1.26
X_2	-2.82	-3.12
X_3	48.69	47.33
X_4	145.05	145.52
X_5	0.06	-0.06
X_6	1.31	0.45
X_7	-0.14	0.50
X_8	0.34	0.37
X_9	0.06	0.10
X_{10}	-0.21	-0.01
X_{11}	0.17	0.02
X_{12}	-0.55	-0.77
X_{13}	-0.08	0.02
X_{14}	0.50	0.88
X_{15}	26.10	46.00
X_{16}	1.86	1.86
X_{17}	-3.24	-3.58

The details of entire analysis are shown in Appendix 3. From Table 10.3, it is clear that there is no error due to misclassification. In other words, the process of selection of normals and abnormals is good. Though the discrimination analysis helps us classify the observations correctly into different groups, we cannot measure the degree of abnormality using this method. Sometimes items or subjects from different groups are encountered according to different probabilities. If we know or can estimate these probabilities a priori, discriminant analysis can use these so-called prior probabilities for assigning the observations to different groups.

The differences between MTS/MTGS and discriminant analysis are as follows:

- In discriminant analysis, the purpose of using Mahalanobis distance is to classify the observation into different groups. Discriminant functions are developed primarily for classifying the observations. They can also be developed when the groups are associated with prior probabilities. The cutoff points can be estimated based on the expected cost of misclassification and prior probabilities.
- In MTS/MTGS the Mahalanobis distance is suitably scaled and used for developing the measurement scale. Only one group (Mahalanobis space) provides the reference point for the scale and correlation matrix for computation of distances. Since abnormals or conditions outside Mahalanobis space are considered unique, the prior probabilities cannot be used in MTS/MTGS methods. The cutoff point (called the threshold) is obtained using quadratic loss function, which minimizes the total cost. A discussion of QLF is provided in Chapter 7.

10.3 STEPWISE REGRESSION

Stepwise regression is widely used for selection of variables in multidimensional systems. The procedure iteratively constructs a sequence of regression models by adding and removing variables at each step. The method requires a specified value of the F-statistic for addition and deletion of the variables in the iterations. The method needs several iterations, if the number of variables is high. A discussion of stepwise regression is given in Montgomery and Peck (1982).

The basic method of stepwise regression is to calculate an F-statistic for each variable in the model. If the model contains $X_1, ..., X_p$, then the F-statistic for X_i is given as

$$F = \frac{SSE\ [X_1, ..., X_{i-1}, X_{i+1}, ..., X_p] - SSE\ [X_1, ..., X_p]}{MSE[X_1, ..., X_{i-1}, X_{i+1}, ..., X_p]}$$

with 1 and $(n - p - 1)$ degrees of freedom. In this equation, the SSE indicates the sum of squares due to errors and MSE indicates mean square error.

If the F-value for any variable is less than F_{out} (specified value) to remove, the variable with the smallest F-value is removed from the model. The regression equation is calculated for this smaller model and the procedure proceeds to a new step. If no variable can be removed, the procedure attempts to add a variable. An F-statistic is calculated for each variable not yet in the model. Suppose the model at this stage contains $X_1, ..., X_p$. Then the F-statistic for a new variable, X_{p+1} is

$$F = \frac{SSE\ [X_1, ..., X_p] - SSE\ [X_1, ..., X_p, X_{p+1}]}{MSE[X_1, ..., X_{i-1}, X_{i+1}, ..., X_p, X_{p+1}]}$$

The variable with the largest F-value is then added, provided its F-value is larger than F_{in} (specified value) to enter. Adding this variable is equivalent to choosing the variable that most effectively reduces the SSE. When no more variables can be entered into or removed from the model, the stepwise procedure ends.

For the medical diagnosis data, stepwise regression analysis is carried out with the help of MINITAB. The results of the analysis are given in Appendix 4. In this table the T-value is an indicator of the F-ratio. After performing stepwise regression analysis, the final model, obtained in the 15th step, is

$$Y = 1.141 - 0.00125X_1 - 0.00114X_5 - 0.0069X_6$$
$$+ 0.0053X_7 + 0.00038X_9 - 0.00177X_{10} - 0.00127X_{11}$$
$$- 0.00191X_{12} + 0.00083X_{13} + 0.00348X_{14}$$
$$+ 0.153X_{15}$$

In other words, X's in the above equation are the important variables based on stepwise regression analysis.

The differences between MTS/MTGS and stepwise regression are as follows:

- Stepwise regression is a probabilistic approach. The method has been criticized because it does not guarantee the best subset regression model. Since the stepwise method ends with one equation, inexperienced analysts may conclude that they have found the optimal subset of the model.
- In MTS/MTGS, we do not require any initial specifications (such as F_{in} and F_{out}). S/N ratios are used to find the useful subset of variables.
- From the results of stepwise regression the important variable combination for the medical diagnosis case is X_1-X_5-X_6-X_7-X_9-X_{10}-X_{11}-X_{12}-X_{13}-X_{14}-X_{15}. The useful variable set obtained by MTS/MTGS analyses with orthogonal arrays is X_2-X_5-X_7-X_{10}-X_{12}-X_{13}-X_{14}-X_{15}.

10.4   TEST OF ADDITIONAL INFORMATION (RAO'S TEST)

The test of additional information proposed by C. R. Rao is also used to identify a set of useful variables. In *Rao's test,* which uses Fischer's linear discrimination function, the test of significance is carried out for a subset of the total number of variables. If the F-ratio is not high, then we can discard the subset of variables. For this test procedure please refer Rao (1973).

The test statistic is calculated using the following equation:

$$F = \left(\frac{N_1 + N_2 - p - q}{p - q}\right) \left(\frac{N_1 N_2 (D_p^2 - D_q^2)}{(N_1 + N_2)(N_1 + N_2 - 2) + N_1 N_2 D_q^2}\right)$$

$$(10.2)$$

where

$D_p^2 = $ the distance with p (all) characteristics
$D_q^2 = $ the distance with a subset q of p characteristics

N_1 = sample size of the first group
N_2 = sample size of the second group

The distance D^2 is calculated with the help of the following equation on the basis of mean vectors of both groups:

$$D^2 = (X_n - X_a)^T C^{-1}_{pooled} (X_n - X_a) \qquad (10.3)$$

where

X_n = mean vector of normal group
X_a = mean vector of abnormal group
C^{-1}_{pooled} = inverse of pooled covariance matrix
$C_{pooled} = \dfrac{1}{N_1 + N_2 - 2} [(N_1 - 1)C_1 + (N_2 - 1)C_2]$
C_1 = covariance matrix corresponding to p variables
C_2 = covariance matrix corresponding to q (a subset of p) variables.

The F-statistic in (10.2) has degrees of freedom $(p - q)$ and $(N_1 + N_2 - p - q)$. A high F-ratio indicates that the variables in the subset provide additional information of the discriminant analysis. Before conducting this test, it is necessary to identify a subset of variables based on some criterion.

For the medical diagnosis data, Rao's test is conducted by considering the subset of variables in Table 10.4. These subsets are same as those selected for multiple Mahalanobis distance method (MMD), discussed in Chapter 9. The details of computations for test 1 (variables in S_1) are given in the following section.

TABLE 10.4

Subset	Variables
S_1	X_1-X_2-X_3-X_4
S_2	X_5-X_6-X_7-X_8
S_3	X_9-X_{10}-X_{11}-X_{12}
S_4	X_{13}-X_{14}-X_{15}-X_{16}-X_{17}
S_5	X_3-X_4-X_5-X_6
S_6	X_{10}-X_{11}-X_{12}-X_{13}-X_{14}-X_{15}
S_7	X_{14}-X_{15}-X_{16}-X_{17}
S_8	X_2-X_5-X_7-X_{10}-X_{12}-X_{13}-X_{14}-X_{15}

Computations for Conducting Test 1

- Subset of variables considered for the test: X_1-X_2-X_3-X_4
- Total number of variables, $p = 17$; number of variables in the subset, $q = 4$
- Sample size of the healthy group, $N_1 = 200$; sample size of the abnormal group, $N_2 = 17$

The pooled covariance matrices for $p = 17$ and $q = 4$ are given in Tables 10.5 and 10.6 respectively. D_p^2 and D_q^2 are computed using Equation 10.3. The values are $D_p^2 = 93.42$ and $D_q^2 = 3.25$. The value of the F-statistic (from Equation 10.2) is 80.10.

In a similar way, Rao's test is performed for other subsets of variables. The summary of the results is given in Table 10.7. From this table, it is clear that the F-ratio is highly significant in all cases (at 5% level of significance), indicating that all subsets are important. This means the F-test may not be adequate to clearly identify the useful set of variables. Thus, though Rao's test provides a means to identify the useful set of variables, it does not guarantee that the chosen set is the optimal one. In addition, we should have prior knowledge on the variables to identify the subset to test.

For all the subsets in Table 10.7, the MTS method is used to compute scaled MDs of abnormals. This is done to determine the ability to measure the degree of abnormality of all subsets. The distinction between these subsets (in terms of scaled MDs) is shown in Figure 10.1. It is clear that the last subset S_8 (the useful subset from the MTS method) has better discrimination and is more consistent than other subsets. The average abnormal MD is also highest for this subset. Interestingly, the F-ratio for this case (Table 10.7) is lower than that of other sets.

10.5 MULTIPLE REGRESSION ANALYSIS

In multiple regression (MR), the objective is to estimate the characteristic y, which is a function of the variables $X_1, X_2, ..., X_k$. If multiple regression is used for the purpose of classification, a decision regarding an observation $(X_1, X_2, ..., X_k)$ can be taken based

TABLE 10.5 Pooled Covariance Matrix (17 Variables)

	X_1	X_2	X_3	X_4	X_5	X_6	X_7	X_8	X_9	X_{10}	X_{11}	X_{12}	X_{13}	X_{14}	X_{15}	X_{16}	X_{17}
X_1	101.82	-11.83	-0.80	-0.70	-174.61	5.39	0.25	27.19	118.38	-4.04	-12.80	48.57	52.21	57.67	-0.85	-0.188	-4.433
X_2	-11.83	18.29	0.10	0.29	242.36	10.20	16.88	-3.70	-16.07	60.77	19.38	-22.89	46.72	-17.18	1.06	2.388	5.534
X_3	-0.80	0.10	0.12	0.04	7.85	-1.63	-1.30	-1.96	0.13	-0.06	0.46	3.17	1.30	1.40	0.01	-0.043	0.059
X_4	-0.70	0.29	0.04	0.05	7.24	-1.19	-0.88	-2.06	-1.03	0.08	0.30	1.43	0.35	0.31	0.02	0.066	0.105
X_5	-174.61	242.36	7.85	7.24	7375.00	-63.97	78.96	-351.74	-290.49	868.55	370.61	246.01	1159.65	89.84	15.99	31.065	89.026
X_6	5.39	10.20	-1.63	-1.18	-63.97	172.00	168.91	189.20	21.96	120.77	40.56	-147.28	84.59	-25.91	0.31	0.010	4.952
X_7	0.25	16.88	-1.30	-0.88	78.96	168.91	214.69	106.96	-24.91	101.21	40.80	-122.39	86.15	-21.13	0.82	-0.303	7.669
X_8	27.19	-3.70	-1.96	-2.06	-351.74	189.20	106.96	641.17	180.93	207.99	81.83	-82.04	153.60	59.44	-0.74	-0.886	1.355
X_9	118.39	-16.07	0.13	-1.03	-290.49	21.96	-24.91	180.93	2053.43	111.82	115.65	121.47	16.62	87.44	-1.41	-11.958	-6.778
X_{10}	-4.04	60.77	-0.06	0.08	868.55	120.77	101.21	207.99	111.82	753.78	220.13	-93.21	280.89	-17.07	3.15	5.108	20.972
X_{11}	-12.80	19.38	0.46	0.30	370.61	40.56	40.80	81.83	115.65	220.13	159.67	33.33	116.07	31.16	1.37	0.620	8.461
X_{12}	48.57	-22.89	3.17	1.43	246.01	-147.28	-122.39	-82.04	121.47	-93.21	33.33	726.63	211.07	498.93	-0.38	3.561	-2.054
X_{13}	52.21	46.72	1.30	0.35	1159.65	84.59	86.15	153.60	16.62	280.89	116.07	211.07	2086.13	275.92	2.50	-0.823	23.087
X_{14}	57.67	-17.18	1.40	0.31	89.84	-25.91	-21.13	59.44	87.44	-17.07	31.16	498.93	275.92	487.65	-0.68	3.114	-0.231
X_{15}	-0.85	1.06	0.01	0.02	15.99	0.31	0.82	-0.74	-1.41	3.15	1.37	-0.38	2.50	-0.69	0.09	0.200	0.356
X_{16}	-0.19	2.39	-0.04	0.07	31.07	0.01	-0.30	-0.89	-11.96	5.11	0.62	3.56	-0.82	3.11	0.20	6.073	0.613
X_{17}	-4.43	5.53	0.06	0.10	89.03	4.95	7.67	1.36	-6.78	20.97	8.46	-2.05	23.09	-0.23	0.36	0.613	2.622

TABLE 10.6 Pooled Covariance Matrix (4 Variables)

	X_1	X_2	X_3	X_4
X_1	101.82	-11.83	-0.80	-0.70
X_2	-11.83	18.29	0.10	0.29
X_3	-0.80	0.10	0.12	0.04
X_4	-0.70	0.30	0.04	0.05

on the value of Y. The multiple regression equation will be of the form

$$Y_i = \beta_0 + \beta_1 X_{i1} + \beta_2 X_{i2} + \cdots + \beta_k X_{ik} + \varepsilon_i$$

where $i = 1,...,n$ (n is the sample size). The constants (β's) are estimated by the method of least squares. The error terms (ε_i's) are assumed to have the following properties:

1. $E(\varepsilon_i) = 0$
2. $V(\varepsilon_i) = \sigma^2$ (constant)
3. $\text{Cov}(\varepsilon_j, \varepsilon_k) = 0, j \neq k$.

A discussion of multiple regression is provided in Montgomery and Peck (1982). For the medical diagnosis data, multiple regression analysis is conducted using MINITAB. A summary of the analysis is given in Appendix 5. The regression equation is of the form

TABLE 10.7 Summary of Rao's Test

Subset	Variables	F-Ratio Calculated	F-Ratio Critical
S_1	X_1-X_2-X_3-X_4	80.1	1.72
S_2	X_5-X_6-X_7-X_8	23.97	1.72
S_3	X_9-X_{10}-X_{11}-X_{12}	47.24	1.72
S_4	X_{13}-X_{14}-X_{15}-X_{16}-X_{17}	33.82	1.75
S_5	X_3-X_4-X_5-X_6	34.59	1.72
S_6	X_{10}-X_{11}-X_{12}-X_{13}-X_{14}-X_{15}	25.4	1.79
S_7	X_{14}-X_{15}-X_{16}-X_{17}	43.26	1.72
S_8	X_2-X_5-X_7-X_{10}-X_{12}-X_{13}-X_{14}-X_{15}	4.12	1.88

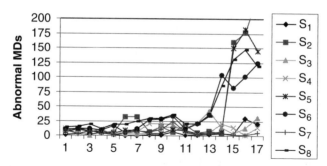

Figure 10.1   Measurement of abnormals for different subsets used in Rao's test.

$$Y = 1.17 - 0.00143X_1 - 0.00273X_2 - 0.0128X_3 + 0.0045X_4$$
$$- 0.00107X_5 - 0.00802X_6 + 0.00596X_7 + 0.000343X_8$$
$$+ 0.000390X_9 + 0.00186X_{10} - 0.00139X_{11} - 0.00203X_{12}$$
$$+ 0.000854X_{13} + 0.00350X_{14} + 0.186X_{15}$$
$$- 0.00006X_{16} - 0.00318X_{17}$$

and the coefficient of determination $R^2 = 0.872$. The high F-ratio obtained in the analysis indicates that the fit is significant. If D_p^2 is the distance between the means of the healthy group and abnormal group, then the relation between D_p^2 and R^2 is given by

$$\frac{R^2}{1 - R^2} = \frac{N_1 N_2 D_p^2}{(N_1 + N_2)(N_1 + N_2 - 2)} \tag{10.4}$$

D_p^2 is computed using Equation 10.4 and is equal to 93.48. This value very closely matches with $D_p^2 = 93.42$ obtained from Equation 10.3.

Thus, multiple regression analysis can be used to classify the observations into different groups based on the regression equation. After ensuring that the fit is good, stepwise regression or Rao's test can be used to identify the variables of importance. The multiple regression equations are developed based on the method of least squares with certain assumptions about the error term. The goodness of the fit is decided based on the F-distribution. If the F-ratio is small, the fit is considered not good. If the fit is not good, then we have to consider other parameters or methods or another set of observations. This is the case even in MTS/MTGS methods if the accuracy of the scale is not good. A disadvantage of multiple regression analysis is that the regression models may become complex if the number of variables is high. Also, it is not easy to measure the degree of abnormality on a continuous scale using the multiple regression method.

10.6 MULTIVARIATE PROCESS CONTROL

When p process variables are being measured simultaneously, multivariate process control charts are used to monitor the process performance over a period of time. There are several types of these charts, such as multivariate Shewhart charts and multivariate Cusum charts. These charts operate just like univariate charts, in which corrective actions are taken whenever the process is out of the control limits.

MTS/MTGS is similar to the process control system in the sense that both of them monitor conditions on a continuous basis. If the conditions are abnormal, necessary corrective actions are taken. In process control charts the degree of abnormality is judged with respect to probabilistic control limits, whereas in MTS/MTGS the degree of abnormality is judged with respect to threshold obtained from QLF.

10.7 ARTIFICIAL NEURAL NETWORKS

Artificial neural networks (ANN) have been in use for pattern recognition, learning, classification, generalization, and interpretation of noisy inputs. ANN can be considered a structure composed of interconnected units (artificial neurons). Each unit has an input/output (I/O) characteristic and implements a local computation or function. The output of any unit is determined by its I/O characteristic, its interconnection to other units. ANN constitutes not one network, but a diverse family of networks.

In this section, MTS/MTGS methods are compared with the feed-forward (backpropagation) method, because this method is commonly used for identifying different patterns. The feed-forward algorithm is one of the methods of backpropagation. The backpropagation method is widely used in the field of classification and pattern recognition. In this method, given inputs and outputs, the method develops a specific nonlinear mapping, which is helpful in diagnostic applications.

10.7.1   Feed-Forward (Backpropagation) Method

This method is primarily used for supervised learning in a feed-forward multilayer perceptron (MLP). Specifically, the back-propagation method is used for training MLPs. Backpropagation was first developed by Werbos in 1974, and is one of the most widely used learning processes in neural networks. Training MLPs with backpropagation algorithms results in a nonlinear mapping or an associated task. Thus, given two sets of data, that is, input /output pairs, the MLP can have its synaptic weights adjusted by the backpropagation algorithm to develop a specific nonlinear mapping. The MLP, with fixed weights after the training process, can provide an association task for classification, pattern recognition, diagnosis, etc. During the training phase of the MLP, the synaptic weights are adjusted to minimize the disparity between the actual and desired outputs of the MLP, averaged over all input patterns.

The standard backpropagation algorithm for training the MLP network is based on the steepest descent gradient method used to minimize the energy function representing the instantaneous error.

Standard Backpropagation Algorithm

1. Initialize the network synaptic weights to small random values.
2. From the set of training input/output pairs, provide input patterns and calculate the network response.
3. Compare the desired network response with actual output of the network and compute all local errors.
4. Update the weights of network.
5. Continue steps 2 through 4 until the network reaches a predetermined level of accuracy in producing the adequate response for all the training patterns.

10.7.2 Theoretical Comparison

MTS/MTGS methods are similar to artificial neural networks in that they do not require any assumptions on the distribution of input variables. But, ANN have certain limitations. They do not provide easy and direct means (as S/N ratios) for dimensionality reduction in terms of the original variables. ANN are primarily used for pattern recognition applications. If a new pattern is added to the system, the weight of the network will be altered. Randomization plays an important role in ANN because as we add new patterns to the system, ANN will tend to forget the old patterns. Therefore, it is necessary to randomize the patterns so that ANN distinguishes all types of patterns in the system. In MTS/MTGS methods if a new pattern is added, the MS corresponding to this pattern has to be constructed and added to the system and there is no need to randomize the patterns. ANN will not provide the relationship between input and output. In MTS/MTGS methods, such relationships can be obtained. Further, in ANN, there is no definite means to set the number of hidden layers, and this number would affect ultimate results.

10.7.3 Medical Diagnosis Data Analysis

The backpropagation method is applied for medical diagnosis data. The network analysis is carried out using MATLAB. At first the network is developed for a training group using the feed-forward method. After that, the network is tested with a test sample. The syntax for feed-forward network analysis is as follows:

$$\text{net} = \text{newff (minmax(input), [S1 S2],}$$
$$\{\text{'tansig' 'tansig'\}, 'traingdx');}$$

Description of the Command

Newff   initializes feed-forward network

Minmax(input)   matrix of min and max values for input elements

S_i size of the ith hidden layer

Tansig   transfer function of the ith layer, default (note that there are other types of transfer functions)

Traingdx   trains a feed-forward network with faster backpropagation and returns an *N*-layer feed-forward backpropagation
network

With this syntax, the network is developed, trained, and tested.
The following commands are used for training and testing the
network:

- *Training*: [net,tr] = train(net,P,T), where train is the function
  for training with input P and output T
- *Testing*: A = sim(net,P), where the function "sim" simulates
  the feed-forward network with input P and correspondingly
  provides the output *A*

For developing and training the network, a sample of 160 (150
normals and 10 abnormals) was selected from the observations of
medical diagnosis data. This sample served as the input file for
the network. The output file consisted of 0's and 1's, with 0's
indicating normals and 1's indicating abnormals. In other words,
the network structure has 17 inputs (17 variables) and output 0 or
1. With this setup, the network has been developed and trained.
The performance goal was set at 0.1 with 500 epochs. Figure 10.2
shows a description of performance goal and number of epochs.
The figure also indicates that the goal was met in 23 epochs.

After training the network, it was tested using a test sample.
The test sample has 57 observations (50 normals and 7 abnormals)
taken from the remaining observations of medical diagnosis data.
From Table 10.8 (output for test data), it is clear that the network
is able to distinguish well between normals and abnormals, since
the normals have values close to 0 and abnormals have values
close to 1. In Table 10.8, column number represents the observations in the test data. The first 50 columns correspond to normals and the remaining 7 columns correspond to abnormals.
Values corresponding to first 50 columns are on the lower side
(somewhat close to 0), indicating that these are normals, and the
values of remaining columns are higher (tending toward 1), indicating that these are abnormals. In other words, the network can
distinguish the normals and abnormals. However, the degree of
abnormality cannot be measured with ANN. The testing stage in

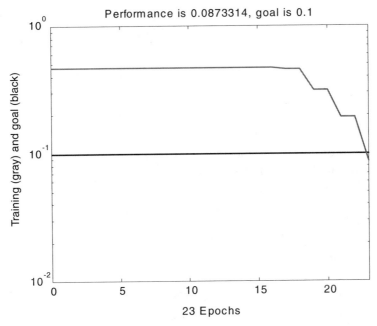

Figure 10.2 Performance goal versus number of epochs (MATLAB output).

TABLE 10.8 MATLAB Output for Test Data

Columns 1 through 7						
0.1818	−0.0386	0.1472	0.0253	0.2754	0.3208	−0.3720
Columns 8 through 14						
0.2614	0.0118	0.4912	0.2290	−0.1059	0.3299	0.2192
Columns 15 through 21						
0.2213	0.1938	0.2769	0.1427	0.1629	0.2264	0.4081
Columns 22 through 28						
0.2099	−0.0233	0.3312	0.2090	0.2813	0.0407	0.1086
Columns 29 through 35						
0.1518	0.3605	−0.0494	0.0992	0.3274	0.3461	0.2328
Columns 36 through 42						
0.0528	0.4984	0.2669	0.1319	0.4601	0.3381	−0.3734
Columns 43 through 49						
0.1322	0.0310	0.2213	0.4409	0.4443	0.0755	−0.2122
Columns 50 through 56						
−0.3009	0.5713	0.6330	0.7978	0.7943	0.9474	0.9171
Column 57						
0.8707						

ANN is similar to that of the validating stage in MTS/MTGS methods. The complete details of MATLAB analysis are given in Appendix 6.

10.8 CONCLUSIONS

- Development of MTS/MTGS methods is based entirely on different thinking and the main purpose is to develop a measurement scale, for multidimensional systems, to carry out effective diagnosis.
- MTS/MTGS methods can be compared with the classical multivariate statistical methods such as PCA, discrimination and classification method, multiple regression, and multivariate process control.
- The objectives of these multivariate/pattern recognition techniques are implicit in MTS/MTGS methods. These are in addition to the primary objective of developing a multidimensional measurement scale.
  - Like PCA, MTS/MTGS methods can be used for dimensionality reduction. In fact, the dimensionality reduction is done in terms of original variables.
  - Like the discrimination and classification method and regression analysis, they can be used to find out the normals and abnormals. Further, MTS/MTGS methods can be used to measure the degree of abnormality.
  - Like stepwise regression or test of additional information, they can be used to screen the variables.
  - Like multivariate charts, they can be used to monitor and control various process conditions.
  - Like ANN, they are distribution free and can be used to recognize various patterns.

11
CASE STUDIES

MTS/MTGS methods can find a wide range of applications in several areas. To demonstrate the potential of these methods, this chapter presents seven case studies from U.S. and Japanese organizations. These case studies are carried out in various fields. The number of MTS-related case studies is increasing every year and at present more than 30 studies have been published.

11.1 AMERICAN CASE STUDIES

11.1.1 Auto Marketing Case Study

This study is related to an auto marketing need to identify the customers' buying pattern of different car segments. This study is part of one of the leadership team projects of the Engineering Management Masters Program (EMMP) for the year 2000. EMMP is a joint program of Wayne State University and Ford Motor Company (Chelst, Falkenburg, and Nagle 1998), by which selected Ford engineers obtain a master's degree. As part of this program students carry out a project in teams. In this study, the objective is to recognize buying patterns of customers owning a particular model, so this is considered a pattern recognition application. The recognition of various patterns using MTS/MTGS analysis can be done as follows:

1. Construct the Mahalanobis space for a pattern under consideration (base pattern).
2. Consider other patterns as abnormals (conditions outside MS).
3. Select the useful variables by using orthogonal arrays and S/N ratios.
4. Use the useful variables for future diagnosis.
5. If we have prior knowledge about the abnormals, then recognize patterns by comparing them with the base pattern. Otherwise, repeat steps 1–4 for all other patterns and test the new observation against all to decide which pattern it belongs to.

The patterns of the buyers were to be identified based on customer survey results. The variables considered for the survey are classified under

1. Personal views
2. Purchase reasons
3. Demographics

The customer survey data were obtained from an "infolink" database. After combining the variables in the above categories, 55 variables were considered (see Appendix 7). The number of car segments is five. The purpose of this case study is to recognize buying patterns of the five segments based on the 55 variables.

In some cases, the customers were asked to rank the variables on a scale of 1 to 4, where 1 means strongly agree and 4 means strongly disagree. The 55 variables were arranged in a desired order and denoted as $X_1, X_2, ..., X_{55}$ for the purpose of analysis. Since there are five segments and we did not have any prior knowledge about these patterns, MTS analysis is performed on all of the segments. For convenience, the five segments are denoted as $S_1, S_2, ..., S_5$.

Construction of Mahalanobis Space

For all of the five segments, the Mahalanobis space is constructed based on a huge data set. For example, MS for S_1 is constructed by taking observations on the 55 variables corresponding to this segment. With these Mahalanobis spaces, the corresponding MDs are calculated.

Validation of the Measurement Scale

The second stage in MTS methods is the validation stage. The outside conditions for a given segment are chosen as conditions corresponding to the other segments. In all of the cases, the abnormals have higher MDs and hence the scale is validated.

Identification of Useful Variables

For this purpose, since we have 55 variables, the L_{64} (2^{63}) orthogonal array (OA) was chosen for analysis. The S/N ratios are computed based on the larger-the-better criterion, because prior information about abnormals was not available.

Table 11.1 provides the list of useful variables corresponding to the five segments under consideration. Since for each segment

TABLE 11.1   Useful Variables Corresponding to the Five Segments

Sample No.	S_1	S/N Ratio Gain	S_2	S/N Ratio Gain	S_3	S/N Ratio Gain	S_4	S/N Ratio Gain	S_5	S/N Ratio Gain
1	X_6	0.20	X_{54}	0.60	X_{52}	1.15	X_7	0.50	X_2	0.96
2	X_{23}	0.16	X_7	0.57	X_{54}	0.73	X_{47}	0.29	X_{47}	0.49
3	X_{15}	0.11	X_{52}	0.48	X_{27}	0.27	X_{31}	0.28	X_{55}	0.49
4	X_{18}	0.11	X_{26}	0.36	X_{24}	0.23	X_{41}	0.27	X_7	0.45
5	X_3	0.11	X_{41}	0.30	X_6	0.19	X_{27}	0.25	X_{40}	0.35
6	X_{52}	0.09	X_{47}	0.28	X_{44}	0.18	X_2	0.22	X_{10}	0.33
7	X_{35}	0.09	X_{25}	0.19	X_3	0.15	X_3	0.22	X_3	0.24
8	X_{19}	0.09	X_{27}	0.16	X_{47}	0.15	X_{21}	0.22	X_{25}	0.20
9	X_{40}	0.09	X_{13}	0.14	X_{25}	0.15	X_{24}	0.16	X_{24}	0.17
10	X_{27}	0.09	X_{18}	0.11	X_4	0.13	X_{22}	0.13	X_{21}	0.16
11	X_{14}	0.08	X_8	0.09	X_2	0.13	X_{15}	0.13	X_6	0.15
12	X_{47}	0.08	X_3	0.08	X_{40}	0.12	X_{54}	0.12	X_{26}	0.15
13	X_{22}	0.07	X_{14}	0.08	X_{26}	0.11	X_8	0.12	X_{18}	0.15
14	X_{12}	0.07	X_{24}	0.08	X_{10}	0.10	X_{23}	0.11	X_{54}	0.14
15	X_5	0.06	X_{40}	0.07	X_{14}	0.10	X_{35}	0.11	X_{29}	0.13
16	X_4	0.06	X_{31}	0.05	X_{42}	0.07	X_{52}	0.10	X_{53}	0.09
17	X_{10}	0.06	X_{29}	0.05	X_{55}	0.06	X_{48}	0.08	X_{49}	0.06
18	X_1	0.06	X_{44}	0.02	X_{41}	0.05	X_{40}	0.07	X_{44}	0.06
19	X_{26}	0.05	X_{48}	0.00	X_{35}	0.05	X_{30}	0.06	X_{35}	0.05
20	X_{11}	0.05			X_{31}	0.01	X_{44}	0.04	X_{42}	0.03

a suitable strategy is to be developed to increase the sales of the cars, it is decided to restrict the number of variables per segment to 20. This is done because the EMMP team felt that it is easier to work with 20 variables and make practicable recommendations. The selection of these variables is done on the basis of the magnitude of gain in S/N ratio. In Table 11.1, in case S_2, the number of useful variables is 19, because these are the only variables with positive gains.

With the useful set of variables, confirmation runs are conducted for all five segments. The results of the confirmation indicate that these variables can recognize the given patterns as effectively as in the case with all the 55 variables. Figure 11.1 shows the recognition power of the useful variables in a given segment.

Table 11.2 gives improvement in the S/N ratios of the entire system for all five segments. The table also provides correspond-

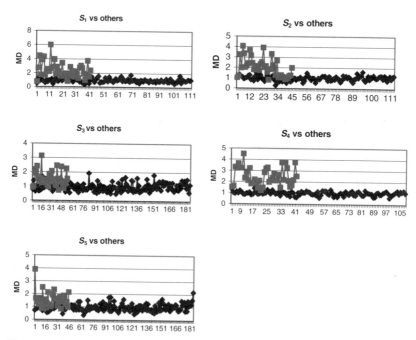

Figure 11.1 Pattern recognition with a useful set of variables. Plots with a higher number of points correspond to respective segments, and plots with fewer points correspond to others (from different segments).

TABLE 11.2   Gain in S/N Ratios and Variability Reduction Range (VRR)

Segment	S/N Ratio Before	S/N Ratio After	Gain	VRR(%)
S_1	2.9	2.98	0.08	0.92
S_2	2.57	2.62	0.05	0.57
S_3	1.12	1.27	0.15	1.72
S_4	1.74	1.79	0.05	0.57
S_5	1.68	1.79	0.11	1.26

ing variability reduction range (VRR). From Table 11.2, it is clear that the gains in S/N ratios are not very significant. This may be because we are dealing with qualitative type of data. However, the reduced number of variables in the optimal system helps reduce the complexity of the multidimensional systems and develop good strategies to improve sales.

11.1.2   Gear-Motor Assembly Case Study

This case study was conducted at Xerox Corporation, a document company. The purpose of life testing the 127K27330 gear-motor assembly is to measure all the parameters (or variables) and decide the reusability of these motors. It is important to identify a suitable measure representing these parameters and based on which a decision on the reusability of a motor can be taken. Since all of these parameters may not be necessary to make a decision, it is necessary to identify a useful set of parameters to determine if a used 127K27330 gear-motor assembly is suitable for reuse. For convenience, the 127K27330 motor is simply referred as "motor" in the remainder of this section. This problem can be treated as a diagnosis problem if reusability of a motor is decided based on the value of the Mahalanobis distance. Hence, MTS/MTGS methods can be applied to this problem.

Apparatus

A diagram of the 127K27330 test fixture can be seen in Figure 11.2. From left to right the holding fixture consists of a hand shield, a high-resolution encoder, a hysteresis brake, customized

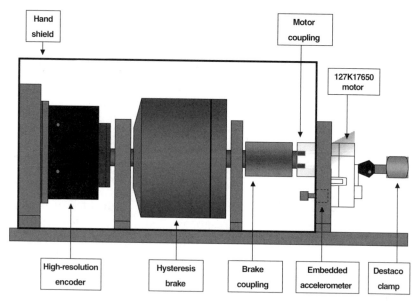

Figure 11.2 The 127K27330 test fixture.

brake coupling, the motor, and a Destaco clamp. The motor's output coupling engages the brake coupling, which turns the hysteresis brake, which turns the encoder. The Destaco clamp prevents any lateral movement.

Sensors

A PCB accelerometer (Model J352A78, 100 mV/g) is used to measure vibration. The accelerometer is stud mounted in the motor mount plate (see Figure 11.2). A 1-Ω resistor is used for current sensing. A Gurley (Model 8435H, 32,000 pulses/rev) high-resolution encoder is used to measure motion and speed.

Detailed Discussion of High-Resolution Encoder

Because the speed of the motor is approximately 1.5 rpm, a standard tachometer could not be used, since its output voltage would be negligible. Instead, a high-resolution encoder from Gurley (Model 8435H) is used for measuring motor speed. This encoder has a resolution of 32,000 pulses/revolution. A U.S. Digital PC6-84-4 quadrature interpolator is used to increase the resolution of this encoder from 32,000 to 128,000 pulses/rev.

A LabView example program titled "Buffered Counting of Events.vi" served as the basis for inputting and interpreting the encoder pulse stream. If, for example, there is an incoming pulse train and one needs to count the pulses and accumulate the count value after every 5 ms, the following needs to be done:

1. Make one of the counters generate a pulse train of 5 ms (200 Hz) and connect the OUT of this counter (No.1) to the GATE of the next counter (No.2).
2. Connect the signal to the SOURCE of counter 2.
3. The pulse train generated by counter 1 goes high every 5 ms.

When it goes high for the first time, the counter 2 will start counting the signal. It keeps counting until it sees another high at the gate, which would be after 5 ms. At this point, the count value of counter 2 is written to the buffer. Counter 2 is actually still counting. When it sees the next high on the gate, after 5 ms, again the counter value is written to the buffer. Therefore, the buffer would have monotonically increasing values representing the accumulated count of counter 1 after each 5-ms interval. We can do this buffered counting at any interval by changing the 200 Hz (5 ms) to some other value.

To summarize, the output of counter 1 is physically tied to the gate of counter 2. Counter 1 time-stamps the running count of counter 2.

Life Test

The life-testing apparatus from the 127K1581 motor test is reused for the 127K27330 life test with a few modifications to the motor base plate. A QuickBasic program titled "dc_motor.bas" and an optomux/opto22 system provided electronic control. Ten new 127K27330 motors are used for the life test. The parameters are measured at intervals of 10,000 cycles starting from 0 cycles to 120,000 cycles during the life test.

Characterization

After every 10,000 cycles the ten motors are taken off the test for characterization. The LabView characterization program is titled "hodaka_dc_motor.vi." The characterization program ran the motor at three different operating conditions: (1) motor voltage (mv) = 21.6 V, brake voltage (bv) = 4.95 V; (2) mv = 21.6 V, bv = 0 V; and (3) mv = 12 V, bv = 4.95 V. These levels are specifically chosen to stress the motor and force a separation between new-motor performance and used-motor performance. The 4.95-V brake voltage corresponds to a 1.96-Nm (276.4-oz-in) load. In these tests, a Compaq personal computer, generic SA interface box and National Instruments AT-MIO-16E-10 board are used for data acquisition.

Of the various parameters measured by hodaka_dc_motor.vi, the following, in the three different operating conditions, are considered to be important:

1. Current
2. Vibration
3. Speed
4. Count–count measurements of counter 2

Some of the parameters cannot be considered, because they are fixed while conducting the life test. Since speed measurements are derived from the counts by taking the derivative of the buffered counts, there is no need to consider both. Therefore, the parameter count is dropped from the analysis. We are left with nine variables: current C, vibration V, and speed S in three operating conditions. The variables are denoted $X_1, X_2, ..., X_9$. The data on these parameters are collected using hodaka_dc_motor.vi.

After collecting data on these nine variables, the MTS method is used for data analysis.

Construction of Mahalanobis Space

Motors 1–10 were all new motors when life testing was started. The presumption is that all new motors are good. Therefore, the parameters measured at 0 cycles (corresponding to these motors) are used to construct the Mahalanobis space for the healthy group. When data of the motors are combined, there are numerous observations corresponding to the parameters at 0 cycles. These parameters reach steady state after some point of time. This can be verified by plotting the data on these parameters on separate graphs. Figure 11.3 shows some of these graphs.

To construct the Mahalanobis space, it is not necessary to consider all of the data. Because the number of observations in the data is large, a subset of the original data is used. The subset selection is done in such a way that it contains an equal number of observations from steady state and from transient state. This helps in constructing a uniform MS. Based on this, MDs corresponding to the observations in the healthy group are calculated.

Validation of MTS Scale

After constructing the measurement scale, it is necessary to test the accuracy of the scale by measuring the MDs of some known

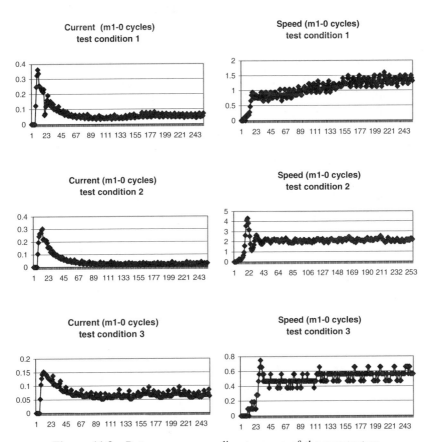

Figure 11.3   Patterns corresponding to some of the parameters.

conditions outside MS. In this application, parameters measured after 10K,20K,....,120K cycles are considered conditions outside MS. Figure 11.4 shows the average MDs corresponding to these cycles. From this figure, we can say that the scale is good because abnormals have higher values of MDs. Note that as the number of cycles increases, the average MD also goes up, indicating that the motor performance deteriorates with an increase in the number of cycles.

Selection of Useful Variables

In the third stage of the analysis, useful parameters are selected using orthogonal arrays and S/N ratios. Since there are nine parameters, the L_{12} (2^{11}) orthogonal array is used. Since we did not have prior knowledge about the abnormals, larger-the-better type S/N ratios are used. The results of MTS analysis are given in Table 11.3. The variable combination X_1-X_2-X_3-X_5-X_6-X_8-X_9 is considered a useful combination because the variables in this combination have positive gains.

A confirmation run is conducted for the useful variables and it is found that from abnormal MDs, we can distinguish normals and abnormals. Figure 11.5 shows the difference between the original combination and the optimal combination in terms of distinction between normals and abnormals. It is clear that the distinction is better with the optimal combination.

Table 11.4 gives the S/N ratio of the entire system before and after optimization. There is a 1.33-dB gain after optimizing the system. This gain is equivalent to a 14% reduction in the varia-

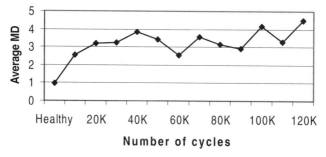

Figure 11.4 Ensuring the accuracy of the scale: number of cycles versus average MD.

TABLE 11.3 *S/N* **Ratio Average Responses and Gains (dB)—MTS Method**

	Level 1	Level 2	Gain
X_1	6.63	5.96	0.67
X_2	7.25	5.34	1.91
X_3	6.58	6.00	0.58
X_4	5.83	6.75	-0.92
X_5	8.03	4.55	3.48
X_6	6.30	6.29	0.01
X_7	6.11	6.48	-0.37
X_8	8.09	4.50	3.59
X_9	7.07	5.51	1.56

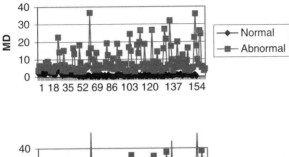

Figure 11.5 Difference between the original combination (top) and the optimal combination (bottom).

TABLE 11.4 *S/N* **Ratio (dB) before and after Optimization**

	S/N Ratio
Before optimization	7.91
After optimization	9.24
Gain	1.33

bility range. Thus, the MDs obtained from useful variables can be used to make decisions about reusability of the motors.

11.1.3 ASQ Research Fellowship Grant Case Study

This case study was carried out under an American Society for Quality (ASQ) research fellowship grant for the year 1999. This case study was used to illustrate the adjoint matrix method, described in Chapter 9.

Following a series of brainstorming sessions, it was decided to apply MTS/MTGS methods to the *process of diagnosis of babies after delivery*. This process was considered critical, as there were several instances of false alarms. Babies weighing between 2500 and 4000 g are considered normal or healthy babies. The existing diagnosis process is based on weight considerations. However, other variables (such as mother's weight and femur length) are also important to making a decision on the condition of babies after delivery. The correlations between the variables also play an important role in identifying the condition of the babies. In view of this, it has been decided to consider the other variables and use MTS/MTGS methods to construct a measurement scale for conducting the diagnosis. The variables for the analysis are selected after a series of discussions with experts in this area. Twelve variables (X_1, X_2, ..., X_{12}) are considered for constructing the measurement scale.

Construction of Mahalanobis Space

As reported in Chapter 9, to construct Mahalanobis space, babies weighing 3000 g $\pm$ 5% are used. Since the correlation matrix (see Table 9.9) has high correlation coefficients and the determinant of this matrix is 8.693×10^{-12}, the MTGS method is used. The useful variables obtained by this method are X_1-X_3-X_4-X_6-X_7-X_{10}-X_{11}-X_{12}. Note that the three highly correlated variables X_{10}, X_{11}, and X_{12} are included in the useful variable list. Because of high correlations, if one variable appears in the useful set, the others also appear in the set. For detailed analysis of this method, please refer Chapter 9.

The next step is to evaluate the gain in S/N ratios after reducing the number of the variables. Table 11.5 gives the summary of

TABLE 11.5   Gain in *S/N* Ratio (dB) of the System

	S/N Ratio
Before optimization	−0.95
After optimization	−0.81
Gain	0.14

S/N ratios before and after optimization. Since the MTGS method can identify the direction of abnormals, it is possible to recognize overweight and underweight babies based on the signs of Gram–Schmidt variables. From Table 11.5 it is clear that there is a 0.14-dB gain in the *S/N* ratio after optimization.

Analysis after Eliminating Two of the Three Correlated Variables

Another way (though not recommended) of dealing with the multicollinearity problem is as follows: If two variables are highly correlated then we can carry out MTS/MTGS analysis with only one of the two variables by discarding the other variable. Therefore, in this case study MTS/MTGS analysis is carried out after eliminating the variables X_{11} and X_{12} from the list of the variables.

As before, the measurement scale is constructed by developing the Mahalanobis space and the accuracy of the scale is ensured. For identifying the useful variable set, an $L_{12}(2^{11})$ orthogonal array is used in MTS. This array is selected because there are 10 variables after discarding X_{11} and X_{12}. Larger-the-better type *S/N* ratios are used to find useful variables. The useful variable combination thus obtained is X_1-X_2-X_3-X_5-X_7-X_9-X_{10}. With this combination, there is a 0.37-dB gain in *S/N* ratio. Refer to Chapter 9 for application of the adjoint matrix method to this problem.

11.1.4   Improving the Transmission Inspection System Using MTS

This case study was conducted at Ford Motor Company. Most of the complex parts in automotives are 100% tested at the end of assembly line, which is called an "end of line" test. These tests are performed using computer-controlled testing machines. The test machine takes 60–120 s to check the functionality of the product before shipment. At Ford, automatic transmissions are typically tested using this procedure. Usually, hundreds of test parameters are measured/calculated and a decision is made based on individual statistical limits on these parameters. The existing method did not take correlations between the variables into account and the method did not provide a means to measure the degree of abnormality.

This case study describes the application of the MTS technique in an end of line test of automatic transmissions. This work is presently ongoing. Because this study made a good impact at Ford for the use of MTS, we decided to report the present status of it. In fact, Ford is trying to develop an "implementation-ready" strategy to use MTS in computerized dynamic testing of complex products, as mentioned above. The initial objective is to use the different stages of MTS for an automatic transmissions system and expand MTS by developing the root cause analysis mechanism.

Application of MTS

1. After conducting a series of discussions for selecting suitable variables for performing MTS analysis, it is decided to consider 392 variables. These variables are selected from a total list of around 900 variables. The 392 variables selected represent different processes performed before transmission assembly. The Mahalanobis space is constructed from a sample of 1900 transmissions. The transmissions that were not returned by the customers are used to construct the healthy group (or Mahalanobis space). Using this space, a measurement scale is developed.

2. To validate the measurement scale, a few known abnormal conditions are selected. The abnormal conditions are selected based on the following criteria:
   • Transmissions returned from the customer
   • Transmissions rejected at end-of-line tests

   For these conditions, the MDs are computed. These MDs indicate that the scale can clearly distinguish conditions outside MS (abnormals).

3. After validating the scale, a screening process is carried out to obtain a useful set of variables. Through larger-the-better S/N ratio analysis, it is found that only 147 variables are important. A confirmation run with the important variables indicates that these variables are sufficient to clearly identify the conditions outside MS. There is a 2.33-dB gain in S/N ratio with the optimal system. This gain is equivalent to 23.6% reduction in variability range.

Current Status

At present, individual diagnosis is being performed to find particular set of variables or processes that are responsible for abnormality. This analysis will help to perform root cause analysis, i.e., concentrate on particular variables or processes and take actions against them to reduce the incidences of abnormality.

11.2 JAPANESE CASE STUDIES

11.2.1 Improvement of the Utility Rate of Nitrogen While Brewing Soy Sauce

This case study describes the use of the MTS method for improving the utility of nitrogen while brewing soy sauce or tamari. Tamari is a special type of thick soy sauce. This study was carried out at Ichibiki & Co. Ltd.

In Japan, 50% of the total requirement of soy sauce and tamari is produced by 5 large-scale makers and the remaining 50% is produced by 1800 small and medium makers. To satisfy Japanese Agricultural standards, soy sauce and tamari makers have consistently produced stable product. Though stable product has been developed, there is a great need to improve brewing technology in the case of small- and medium-scale manufacturers to compete with the technology of large-scale manufacturers and to maintain consistent quality and market share. In view of this, it has been decided to use MTS methodology to improve the brewing process. Since nitrogen content contributes significantly to the deliciousness of soy sauce, it is important to increase the utility rate of nitrogen during the brewing process. The utility rate of nitrogen is defined as the proportion of nitrogen that has dissolved in Moromi (unrefined soy sauce). The upper limit of the utility rate of nitrogen was fixed at 92% based on nitrogen that will not dissolve.

Process of Producing Soy Sauce or Tamari

There are four subprocesses in the production of soy sauce or tamari:

1. Material treatment
2. Preparation of koji molds (special molds for brewage)
3. Aging
4. Inspection

The contributing factors in these subprocesses are as follows:

Material Treatment   Type of materials, compounding, amount of water spraying, season for treatment, salt solution, condition of steaming and boiling, and so on

Preparation of Koji Molds   Chamber of koji molds, amount to be added, compounding, setting condition of air conditioner, temperature, and so on

Aging   Temperature, agitation, condition of koji molds

Inspection   Number of koji mold bacteria, water content, power factor of mold–protein decomposing enzyme (PU30), pH, color, sediment, salt content, total nitrogen, alcohol, power factor of protein decomposing enzyme (PU15) in Moromi, utility rate of dissolved nitrogen in Moromi.

Note: Testing methods shall conform to the specifications developed by Nihon Research Laboratory for soy sauce.

Selection of Factors for MTS Application

As mentioned before, the deliciousness of soy sauce significantly depends on the nitrogen content. Soy sauce contains 1.5–1.7% nitrogen and 2–3% alcohol, while tamari contains 2–3% nitrogen and 0.5% alcohol. Since manufacturing factors of soy sauce and tamari are common, the corresponding data are not treated separately; rather, they are combined. During the production of soy sauce or tamari, the subprocesses koji mold preparation and aging are critical, and hence the MTS method is applied to these subprocesses separately. The number of factors for koji mold is 40 and that for aging is 82. The historical data for these factors was collected from the years 1996 to 1999.

MTS for Aging

The Mahalanobis space is generated based on the upper limit on the utility rate of nitrogen and on 203 observations of 82 factors. Using MS, the measurement scale is developed with the help of MDs of normals. After this step, the scale is validated with known conditions. After performing dynamic S/N ratio analysis, the number of factors is reduced from 82 to 9. These factors are f2, K1ge, k5, pu15, pH5end, iro3, roux, S, al. Factor pu15, which is the power factor of the protein-decomposing enzyme in Moromi resulting from koji molds, is an important factor for improving the preparation process of koji molds.

MTS for Koji Molding

Factor pu 15 in koji molds is correlated with water content in koji molds. As water content increases, the pu 15 will also be high. Since the measuring water content is easier than measuring of pu 15, the MS was constructed based on the water content. The data for MS are obtained from 40 factors in a sample of 614. After validating the measurement scale, dynamic S/N ratio analysis was conducted to identify a useful set of factors. The number of useful factors, in this case, is found to be 8. These factors are rr, muro, m.wash, kisetu, mori.h, c.up.h, c.up.ond, t2.10h.

The useful factors obtained through MTS are sufficient to measure the quality of soy sauce or tamari and take appropriate actions to maintain the required utility rate of nitrogen. This method of MTS analysis is included in the ISO 9001 quality system of this company.

11.2.2 Application of MTS for Measuring Oil in Water Emulsion

This study describes use of MTS to predict the "healthiness" of oil in water emulsion. The study was conducted at Fuji Film Co.

Ltd. Sensitive materials such as negative color film consist of very thin sensitization layers. They make use of various other materials and each provides a specific function to satisfy the requirements of sensitive materials.

The sensitization layers use gelatin as binder, and a water-soluble material can be directly added to sensitive materials as an aqueous solution. An oil-soluble organic material, which is soluble in a solvent, can be added to sensitive materials with drops of oil (oil in water emulsion) by using surface active agents. Whether oil-soluble material provides a function in sensitive materials depends on the properties of drops of oil. The precipitation of materials inside drops of oil or rise in grain size of drops of oil by coalescence may result in a reduced performance of the materials. Therefore, it is necessary to optimize type, quantity of many materials, solvent and surface active agents to design drops of oil so that precipitation or coalescence does not occur.

To measure the conditions of drops of oil and the variables affecting them, MTS methodology is used. The useful variables should be able to predict the conditions of precipitation and coalescence. Figure 11.6 shows typical conditions of drops of oil. The different conditions of drops of oil are referred to as recipes.

Application of MTS

After defining the normal and abnormal conditions, the MTS method is systematically applied. The Mahalanobis space is constructed from the historical data corresponding to the recipes hav-

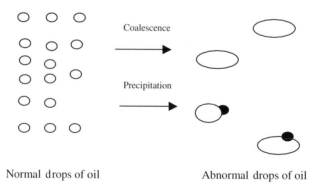

Figure 11.6 Drops of oil.

ing no problems of precipitation and coalescence. The variables defining these recipes include various additives, solvents, and surface active agents. From MS, the MDs are computed for normals. For known abnormals, MDs are computed to validate the measurement scale. Here, the abnormal conditions are generated by carrying out experiments. It was found that the normal recipes have MDs of 5.0 or less and abnormals have very high MDs. For recipes with MDs of 5.0 or less, the number of incidences of occurrence of coalescence or precipitation is very low. Thus, for these recipes, the need for conducting compulsive experiments is greatly reduced, which saves man-hours and cost. This shows the power of MD to predict failures while minimizing total cost.

In the next step, the useful variables are identified with help of S/N ratios to predict failures in a more efficient manner. After developing the MTS scale with useful variables, it is applied to actual recipe designs. The MDs corresponding to these recipes are able to predict the conditions accurately. This was ascertained by carrying out detailed experimentation.

Thus, the MTS method is very helpful in predicting abnormal recipes so that appropriate corrective actions can be taken. This helps in using the right kind of recipe so that the performance of sensitive materials is not affected, which reduces unnecessary experimentation and man-hours.

11.2.3 Prediction of Fasting Plasma Glucose (FPG) from Repetitive Annual Health Checkup Data

This case study describes use of the MTS technique for predicting fasting plasma glucose (FPG) to control diabetes mellitus. The study was conducted by Aichi Health Plaza, Health Science Museum, and Nakaroku Ltd. In a survey conducted by the Japanese Ministry of Health and Welfare in 1997, it was reported that about 6.9 million Japanese have diabetes and about 13.7 million Japanese have blood glucose content different from the standard level. The blood glucose content can be used as a measure of diabetes mellitus. Diabetes mellitus largely depends on day-to-day activities (like food and exercise habits). When detected at early stage, diabetes mellitus can be mitigated by improving day-to-day activities. Because diabetes mellitus depends on several variables, the

MTS method is used to predict this disease through FPG. The prediction results are also compared with stepwise regression analysis.

Diabetes Mellitus

Diabetes mellitus is a group of diseases characterized by chronic hyperglycemia based on the degree of deficiency of insulin effect, decrease in insulation secretion, metabolic abnormalities, and chronic hyperglycemia. The Japanese Diabetes Society provided the following standard criterion to make a decision on these diseases:

Diabetes   FPG ≥ 126 mg/dL
Normal   FPG < 110 mg/dL
Border   those who are in between diabetes and normal

For diagnosing these diseases, one has to ascertain FPG values at least two times.

Application of MTS

The normal and abnormal conditions are defined in accordance with standards of the Diabetes Society described above. Various factors, such as age, diabetes family history, food habits, exercise habits, corpulence degree (BMI), and blood pressure, are considered for MTS analysis.

Initially data on these variables was collected for one year. To improve the accuracy of prediction, the number of observations was increased by adding subsequent years of data. This process was stopped after collecting five years data. From these data, Mahalanobis space was constructed. After validating the measure-

TABLE 11.6   Predictions Based on MTS and Stepwise Regression

	FPG Value (mg/dl)
MTS-based prediction	134.3 ± 7.2
Prediction from stepwise regression	117.7 ± 6.2
Measured value	132.7 ± 9.0

ment scale, *S/N* ratio analysis was conducted to find the useful set of variables. It was found that this useful variable set, thus obtained, is sufficient for future predictions of FPG (and hence diabetes mellitus) based on MDs. As shown in Table 11.6, the predictions based on MD are better than those with stepwise regression analysis. In other words, MTS based predictions are much closer to the measured values than predictions based on stepwise regression analysis. Thus, the MTS method is very useful in predicting diabetes mellitus, which will enable patients to take appropriate preventive actions.

11.3 CONCLUSIONS

- Through the case studies discussed in this chapter, it is clear that MTS/MTGS methods have great potential in several fields.
- The cases also demonstrate the use of the proposed methods in situations like multi-collinearity.
- These cases also show that MTS-based predictions match very closely with measured values.
- Encouraged by initial success, some industries are planning to expand the application of MTS in different areas.

12
CONCLUDING REMARKS

This chapter highlights the important points of the MTS/MTSG methods and their scientific contributions for practical adaptations with success. The chapter also provides limitations of these methods and recommendations for future research.

12.1 IMPORTANT POINTS OF THE PROPOSED METHODS

1. Application of MTS/MTGS methods can be considered a subject of measurement science for multidimensional systems, in which a measurement scale is developed. Using this scale, severities of different conditions are measured.

2. The reference point for the measurement scale can be obtained by constructing the Mahalanobis space (MS). MS is a database consisting of means, SDs, and the correlation structure of the variables in the healthy group. The development of MS depends on the choice of the specialist conducting the diagnosis.

3. The definition of a healthy group (MS) is extremely important, because MDs corresponding to observations in this group represent different patterns. Since MDs depend on correlation structure and we cannot calculate correlations

with one observation, we need a group (a healthy group) of observations. Based on this correlation structure, MDs (patterns) can be obtained. It is important that the healthy group be as uniform as possible. The observations in the healthy group (MS) are considered best for a particular scale to which it provides the zero point and unit distance.

4. Because MD is used only to measure the distances in multidimensional systems, we are not interested in its distribution. In Chapter 2 it was proved that we do not require any distribution to specify MDs in MS and irrespective of the distribution of the input variables, MD can be used to define the zero point and unit distance.

5. MTS and MTGS are two different methods: each method has advantages and disadvantages over the other. A detailed comparison between these methods is provided in Chapter 3.

6. As in robust engineering (or Taguchi Methods), the S/N ratio plays an important role in MTS/MTGS. This robust measure has several advantages. The role of S/N ratios in MTS/MTGS along with their advantages is clearly described in Chapter 4.

7. One of the important features of Taguchi Methods is the use of the quadratic loss function (QLF) to define specifications for the parameters of a product. In MTS/MTGS, QLF plays an important role in providing the threshold for the measurement scale. The MDs are compared with the threshold to determine the severity of the conditions. Based on the degree of severity, appropriate corrective actions can be taken. A discussion of QLF in relation to MTS/MTGS is provided in Chapter 7.

8. MTS/MTGS methods can be effectively applied in situations having categorical data. This aspect is illustrated with a case study in Chapter 5.

9. MTS/MTGS methods can be handled even in the presence of noise factors. The noise factors could be different hospitals or different inspection systems. There are different ways of using these methods in the presence of noise factors. Chapter 6 describes in detail the different ways of using MTS/MTGS in the presence of noise factors.

10. The adjoint matrix method can safely replace the inverse matrix method (MTS), since it is as efficient as the inverse matrix method in general and more efficient when there are problems of multicollinearity (strong correlations). However, MTGS is required to identify the direction of abnormals.

11. Use of the β-adjustment method is recommended when there are small correlation coefficients or weak correlation between variables.

12. Use of the MMD method will reduce the problem complexity and help one make effective decisions in larger problems. In the MMD method, the large number of variables is divided into several subsets containing local variables. Through S/N ratio analysis of MMDs, useful subsets are identified.

13. Chapter 9 provides a detailed discussion on the adjoint matrix method, β-adjustment method, and MMD method. Chapter 9 also gives guidelines for selecting a good MS from historical data by which "healthiness" of a condition cannot be defined.

14. MTS/MTGS methods are much different from other multivariate/pattern recognition methods. Chapter 10 provides a detailed discussion of the comparison between MTS/MTGS methods and classical methods including artificial neural networks. The main objective in MTS/MTGS is to construct a measurement scale for multivariate systems. Like ANN, these methods are data analytic and are not developed based on any probability distribution. The current trends in multivariate diagnosis/pattern recognition lean toward the data analytic procedures.

15. MTS/MTGS can find applications in several different areas. The case studies discussed in the book show the potential of these methods in different areas of application.

12.2   SCIENTIFIC CONTRIBUTIONS FROM MTS/MTGS METHODS

1. MTS/MTGS methods provide a methodology for developing a measurement scale for multidimensional systems.

2. The proposed methods, discussed in this book, constitute a new subject for the measurement science in multidimensional information systems.

3. The methods integrate mathematical and statistical concepts, such as the Mahalanobis distance and the Gram–Schmidt orthogonalization method, with the principles of Taguchi Methods.

4. The current trends in multivariate diagnosis/pattern recognition call for solution methods that are not based on distributions or probabilities. MTS/MTGS methods are developed to satisfy the current trends.

5. The objectives of many existing multivariate/pattern recognition techniques are implicit in MTS/MTGS methods. These are in addition to the primary objective of developing a multidimensional measurement scale

6. Use of signal-to-noise ratios (S/N ratios) is highly recommended in multidimensional applications. S/N ratios can be used to reduce the dimensionality of the system in terms of the original variables, to measure functionality of the system, and to predict the given conditions.

7. A procedure is outlined to set thresholds by minimizing the total cost. This helps in reducing false alarms, time, and effort.

8. The adjoint matrix method, MMD method, and β-adjustment methods are useful in handling difficult problems that are frequently encountered in multivariate/pattern recognition applications.

9. The development of the measurement scale and diagnosis process can be considered development of expert systems using Mahalanobis distance. By using this method, a methodology can be formulated to set up different zones for the treatment of different conditions based on their severity and associated cost.

10. Use of MTGS method helps to determine the direction of abnormals.

11. The number of MTS-related case studies is increasing every year, and at present there are more than 30 published case studies.

12.3 LIMITATIONS OF THE PROPOSED METHODS

1. Definition of Mahalanobis space depends on the discretion of the decision maker. The Mahalanobis space can be constructed by defining the "healthiness" of a condition or from historical data. If there is something wrong in the decision maker's judgment, then the scale gives wrong measurements and hence cannot be used for its intended purpose.

2. MTGS cannot be used to identify the direction of abnormals, where knowledge about the type of variables (like larger-the-better type or smaller-the-better type) is not available.

3. Use of MTGS without orthogonal arrays is not recommended for dimensionality reduction if partial correlations are significant and there does not exist a definite order of the variables.

4. There is no specific set of guidelines to choose a particular method of treating noise factors in proposed methods.

12.4 RECOMMENDATIONS FOR FUTURE RESEARCH

1. There is a need to extend these methods for carrying out root-cause analysis. A methodology for this purpose should be developed.

2. These methods should be used in larger problems related to voice recognition, pattern recognition and economics.

BIBLIOGRAPHY

Agnew, J. L., and Knapp, R. C. (1988). *Linear Algebra with Applications.* Pacific Grove, CA: Brooks–Cole.

Algorithm: Mahalanobis distance. *http://www.galactic.com/galactic/science/maldist.html*

Anderson, T. W. (1984). *An Introduction to Multivariate Statistical Analysis,* 2nd ed. New York: Wiley.

ASI (1998). *Robust Design Using Taguchi Methods.* An ASI Workshop Manual. Dearborn, MI: ASI Press.

Bentley, D. L., and Cooke, K. (1974). *Linear Algebra with Differential Equations.* New York: Holt, Rinehart & Winston.

Bewig, K. M., Clarke, A. D., Roberts, C., and Unlesbay, N. (1994). Discriminant analysis of vegetable oils by near-infrared reflectance spectroscopy. *Journal of the American Oil Chemists' Society,* **71**(2): 195–200.

Brown, W. C. (1991). *Matrices and Vector Spaces.* New York: Decker.

Cao, X.-R., and Zhou, Z. (1993). Correlation analysis in quality control of the lift gate assembly line. *Quality Engineering,* **5**(4): 601–609.

Chelst, K., Falkenburg, D., and Nagle, D. (1998). An industry-based engineering management masters program for the working engineer. *Journal of Engineering Education,* **87**(3): 289–297.

Clausing, D. (1994). *Total Quality Development.* New York: ASME.

Dasgupta, S. (1993). The evolution of the D^2-statistic of Mahalanobis. *Sankhyā,* (Series A), **55,** Part 3: 442–459.

Dillon, W. R., and Goldstein, M. (1984). *Multivariate Analysis: Methods and Applications.* New York: Wiley.

Duda, R. O., and Hart, P. E. (1973). *Pattern Classification and Scene Analysis.* New York: Wiley.

209

Ealey, L. A. (1994). *Quality by Design: Taguchi Methods and U. S. Industry.* Dearborn, MI: ASI Press.

Fowlkes, W. Y., and Creveling M. C. (1995). *Engineering Methods for Robust Product Design: Using Taguchi Methods in Technology and Product Development.* Reading, MA: Addison-Wesley.

Furutani, H., Kitazoe, Y., Yamamoto, K., Ogura, H., et al. (1984). Evaluation of the Mahalanobis generalized distance—method of quality control: monitoring system of multivariate data. *American Journal of Clinical Pathology,* **81**(3): 329–336.

Gerth, R. J., and Hancock, W. M. (1995–96). Output variation reduction of production systems involving a large number of subprocesses. *Quality Engineering,* **8**(1): 145–163.

Ham, F. M., and Kostanic, I. (2001). *Principles of Neurocomputing for Science & Engineering.* New York: McGraw-Hill.

Hancock, D. O., and Synovec, R. E. (1989). Early detection of C-130 aircraft engine malfunction by principal component analysis of the wear metals in C-130 engine oil. *Applied Spectroscopy,* **42**(2): 202–208.

Hancock, W. M., Yoon, J. W., and Plont, R. (1996). Use of Ridge regression in the improved control of casting processes. *Quality Engineering,* **8**(3): 395–403.

Hasegawa, Y. (1997). How to treat missing data in health examination using Mahalanobis distance. *ASI's Total Product Development Symposium Proceedings,* 619–630.

Hassoun, M. H. (1995). *Fundamentals of Artificial Neural Networks.* Cambridge, MA: MIT.

Hawkins, D. M. (1991). Multivariate quality control based on regression-adjusted variables. *Technometrics,* **33**(1): 61–76.

Hohn, F. E. (1967). *Elementary Matrix Algebra.* New York: Macmillan.

Holmes, D. S., and Mergan, A. E. (1995–96). Identifying the sources for out-of-control signals when the T^2 control chart is used. *Quality Engineering,* **8**(1): 137–143.

Hu, S. J., and Wu, S. M. (1992). Identifying sources of variation in automobile assembly using principal component analysis. *NAMRI/SME Transactions,* **XX**: 311–316.

Hu, S. J., Wu, S. K., and Wu, S. M. (1991). Multivariate analysis and variation reduction case studies in automobile assembly. *NAMRI/SME Transactions,* **XIX**: 303–308.

Huhen, J. H., Hollon, K. R., and Lai, D. C. (1990). Comparison of Mahalanobis distance, polynomial and neural net classifiers. *SPIE Proceedings— Applications of Artificial Neural Networks,* **1294**: 557–567.

Jackson, J. E. (1985). Multivariate quality control. *Communication in Statistics—Theory and Methods,* **14**(11):2657–2688.

Johnson, R. A., and Wichern, D. W. (1992). *Applied Multivariate Statistical Analysis.* Englewood Cliffs, NJ: Prentice Hall.

Jugulum, R. (2000). *New Dimensions in Multivariate Diagnosis to Facilitate Decision Making Process.* Thesis, Wayne State University, Detroit, MI.

Jugulum, R., and Monplaisir, L. (2002). Comparison between Mahalanobis–Taguchi system and artificial neural networks. *Journal of the Japanese Quality Engineering Society,* **10**(1):60–73.

Jugulum, R., and Taguchi, S. (2001). Direction of abnormals and treatment of noise factors in multivariate diagosis: a discussion. *Journal of the Japanese Quality Engineering Society,* **9**(3): 37–44.

Jugulum, R., Taguchi, S., and Yang, K. (1999). New developments in multivariate diagnosis: a comparison between two methods. *Journal of the Japanese Quality Engineering Society,* **7**(5): 62–72.

Kamoshita, T. (1997). Optimization of a multi-dimensional information system using Mahalanobis distance. *ASI's Total Product Development Symposium Proceedings,* 675–687.

Kanetaka, T. (1988). Application of Mahalanobis distance, standardization and quality control. *Japanese Standards Association,* **41**(5 and 6).

Kang, J., Matsumura, Y., and Mori, T. (1991). Characterization of texture and mechanical properties of heat-induced soy protein gels. *Journal of the American Oil Chemists' Society,* **68**(5): 339–345.

Mahalanobis, P. C. (1936). On the generalized distance in statistics. *Proceedings, National Institute of Science of India,* **2**: 49–55.

Mahalanobis, P. C., and Bose, C. (1943). Correlation between anthropometric characters in some Bengal castes and tribes. *Sankhyā,* **5**: 249–260.

Mason, R. L., Tracy, N. D., and Young, J. C. (1995). Decomposition of T^2 for multivariate control chart interpretation. *Journal of Quality Technology,* **27**(2): 99–108.

Montgomery, D. C. (1984). *Design and Analysis of Experiments.* New York: Wiley.

Montgomery, D. C., and Mastrangelo, C. M. (1991). Some statistical process control methods for autocorrelated data. *Journal of Quality Technology,* **23**(3): 179–193.

Montgomery, D. C., and Peck, E. (1982). *Introduction to Linear Regression Analysis.* New York: Wiley.

Morrison, D. F. (1967). *Multivariate Statistical Methods.* New York: McGraw-Hill.

Morrison, D. F. (1990). *Multivariate Statistical Methods,* 3rd ed., McGraw-Hill Series in Probability and Statistics. New York: McGraw-Hill.

Nair, V. N. (1993). Taguchi's parameter design: a panel discussion. *Technometrics,* **34**(2): 127–161.

Park, S. H. (1996). *Robust Design and Analysis for Quality Engineering.* New York: Chapman & Hall.

Perlis, S. (1958). *Theory of Matrices.* Reading, MA: Addison–Wesley.

Phadke, M. S. (1989). *Quality Engineering Using Robust Design.* Englewood Cliffs, NJ: Prentice Hall.

Phadke, M. S., and Taguchi, G. (1987). Selection of quality characteristics and S/N ratios for robust design. *Proceedings of the IEEE GLOBECOM-87 Conference, Tokyo, Japan,* 1002–1007.

Rao, C. R. (1973). *Linear Statistical Inference and Its Applications.* New York: Wiley.

Sato, T. (1994). Application of principal component analysis on near-infrared spectroscopic data of vegetable oils for their classification. *Journal of the American Oil Chemists' Society,* **71**(3): 293–298.

Taguchi, G. (1987). *System of Experimental Design,* Vols. 1 and 2. Dearborn, MI: ASI Press.

Taguchi, G. (1993). *Taguchi on Robust Technology Development.* New York: ASME.

Taguchi, G. (1994). Diagnosis and signal-to-noise ratio. *Quality Engineering Forum,* **2** (4 and 5).

Taguchi, G. (1994). Application of Mahalanobis distance for medical treatment. *Quality Engineering Forum,* **2**(6).

Taguchi, G. (1994). *Taguchi Methods,* Vol. 1: *Research & Development;* Vol. 2: *On-Line Production;* and Vol. 4: *Design of Experiments.* Tokyo: Japanese Standards Association and Dearborn, MI: ASI press.

Taguchi, G. (1995). Multivariate sensing system design. *Quality Engineering Forum,* **3**(1).

Taguchi, G., and Konishi, S. (1993). *Taguchi Methods: Design of Experiments.* Dearborn, MI: ASI Press.

Taguchi, G., and Jugulum, R. (1999). Role of S/N ratios in multivariate diagnosis. *Journal of the Japanese Quality Engineering Society,* **7**(6): 63–69.

Taguchi, G., and Jugulum, R. (2000). Quality loss function in multivariate diagnosis. *Journal of the Japanese Quality Engineering Society,* **8**(4): 47–52.

Taguchi, G., and Jugulum, R. (2000). New trends in multivariate diagnosis. *Sankhyā* (Series B) **62,** Part 2: 233–248.

Tracy, N. D., Young, J. C., and Mason, R. L. (1992). Multivariate control charts for individual observations. *Journal of Quality Technology,* **24**(2): 88–95.

Wu, Y. (1996). Pattern recognition using Mahalanobis distance. *ASI's Total Product Development Symposium Proceedings,* 212–218.

APPENDIXES

TABLE A.1 ASI Data Set

	X_1	X_2	X_3	X_4	X_5	X_6	X_7	X_8	X_9	X_{10}	X_{11}	X_{12}	X_{13}
1	59.2	69.5	73.4	81.5	88.1	104.3	118.4	140.5	151.5	156.6	168.4	181.7	193.3
2	59.1	69.3	73.3	81.2	88.0	104.2	118.0	140.0	150.9	156.2	168.2	181.3	192.5
3	59.2	69.5	73.4	81.2	88.0	104.5	118.2	140.1	151.2	156.2	168.4	181.5	192.7
4	59.2	69.6	73.4	81.5	88.1	104.6	118.5	140.4	151.3	156.7	168.9	181.9	193.4
5	58.9	69.2	73.1	81.3	87.7	103.7	118.0	139.8	150.8	156.1	167.6	180.9	192.5
6	59.1	69.8	73.3	81.2	87.9	104.8	118.6	140.1	151.3	156.2	169.1	181.6	193.4
7	59.1	69.5	73.4	81.3	88.0	104.3	118.3	140.1	151.4	156.3	168.4	181.4	193.1
8	59.4	70.2	73.6	82.2	88.4	105.1	119.7	141.3	152.6	157.6	169.8	183.0	195.3
9	59.0	69.3	73.2	81.0	87.8	104.2	117.9	139.8	150.7	155.8	168.1	180.7	192.3
10	59.2	69.6	73.3	81.6	88.0	104.5	118.7	140.4	151.5	156.7	168.8	181.8	193.6
11	59.2	69.6	73.4	81.4	88.0	104.5	118.6	140.3	151.6	156.5	168.7	181.7	193.5
12	59.2	69.7	73.5	81.8	88.1	104.6	118.9	140.6	151.7	156.9	169.0	182.0	193.9
13	59.4	70.5	73.8	82.1	88.5	105.5	119.9	141.3	152.9	157.6	170.4	183.2	195.6
14	59.2	69.7	73.5	81.5	88.1	104.7	118.6	140.5	151.4	156.6	169.1	182.0	193.4
15	59.3	69.5	73.6	81.3	88.2	104.7	118.1	140.3	151.1	156.5	168.9	181.9	192.7
16	59.1	69.6	73.3	81.3	88.0	104.6	118.4	140.2	151.3	156.3	168.9	181.6	193.2
17	59.3	69.7	73.5	81.5	88.2	104.8	118.5	140.5	151.5	156.7	169.2	182.1	193.5
18	59.1	69.4	73.2	81.1	87.9	104.2	117.9	139.9	150.8	156.0	168.2	181.3	192.4
19	59.2	69.7	73.5	81.5	88.1	104.6	118.6	140.5	151.6	156.7	168.9	181.8	193.6
20	59.4	69.7	73.6	81.5	88.4	104.8	118.5	140.7	151.6	156.9	169.0	182.2	193.3
21	59.4	69.7	73.6	81.5	88.3	104.8	118.5	140.6	151.5	156.7	169.0	182.0	193.3
22	59.3	69.7	73.5	81.5	88.2	104.6	118.7	140.5	151.6	156.7	168.8	181.9	193.5
23	59.3	69.6	73.5	81.5	88.2	104.5	118.5	140.5	151.6	156.8	168.7	182.0	193.3
24	59.3	69.6	73.6	81.4	88.2	104.6	118.5	140.4	151.4	156.6	168.8	181.8	193.2
25	59.3	69.6	73.6	81.4	88.2	104.6	118.5	140.4	151.6	156.7	168.9	181.8	193.4

(continues)

TABLE A.1 (*Continued*)

	X_1	X_2	X_3	X_4	X_5	X_6	X_7	X_8	X_9	X_{10}	X_{11}	X_{12}	X_{13}
26	59.2	69.5	73.4	81.4	88.1	104.4	118.0	140.2	150.8	156.5	168.6	181.7	192.7
27	59.3	69.6	73.5	81.5	88.2	104.7	118.5	140.5	151.3	156.7	169.0	181.9	193.2
28	59.2	69.4	73.3	81.4	88.0	104.4	118.1	140.2	150.9	156.5	168.6	181.6	192.7
29	59.3	69.7	73.6	81.7	88.2	104.7	118.6	140.7	151.5	156.9	169.1	182.1	193.6
30	59.3	69.6	73.5	81.7	88.2	104.6	118.6	140.7	151.3	156.9	168.9	182.1	193.5
31	59.2	69.6	73.3	81.3	88.0	104.3	118.3	140.2	151.2	156.2	168.3	181.5	192.9
32	59.3	69.5	73.5	81.7	88.2	104.4	118.3	140.7	151.3	156.9	168.6	181.9	193.3
33	59.2	69.8	73.4	81.6	88.1	104.5	118.8	140.4	151.8	156.7	168.7	181.8	193.7
34	59.2	69.4	73.4	81.4	88.0	104.3	118.2	140.3	151.2	156.4	168.4	181.5	192.9
35	59.4	70.0	73.6	81.8	88.4	105.0	119.3	140.9	152.3	157.2	169.4	182.5	194.5
36	59.3	69.7	73.5	81.6	88.2	104.6	118.7	140.5	151.7	156.8	168.9	181.9	193.5
37	59.1	69.5	73.3	81.5	88.0	104.3	118.5	140.3	151.2	156.6	168.6	181.5	193.2
38	59.1	69.5	73.3	81.4	87.9	104.4	118.3	140.1	151.0	156.4	168.6	181.4	193.0
39	59.1	69.5	73.3	81.4	88.0	104.5	118.2	140.2	150.9	156.4	168.6	181.5	192.8
40	59.5	70.2	73.8	81.8	88.6	105.4	119.5	141.1	152.7	157.4	170.1	183.0	194.9
41	59.2	69.6	73.4	81.5	88.0	104.4	118.7	140.4	151.7	156.6	168.5	181.7	193.5
42	59.6	70.3	73.8	82.2	88.6	105.5	119.9	141.5	152.9	157.7	170.3	183.3	195.4
43	59.3	69.9	73.6	81.9	88.2	104.9	119.2	140.9	151.9	157.1	169.5	182.4	194.4
44	59.3	69.8	73.5	81.6	88.2	104.7	118.8	140.6	151.7	156.8	169.1	182.1	193.8

TABLE A.2 Principal Component Analysis (MINITAB Output)

Variable	PC1	PC2	PC3	PC4	PC5	PC6
Z1	-0.166	-0.256	-0.135	-0.200	-0.402	-0.420
Z2	-0.131	0.453	0.090	-0.084	-0.090	-0.214
Z3	0.169	0.181	-0.362	0.302	0.196	0.210
Z4	0.222	0.298	-0.257	0.074	0.073	0.100
Z5	0.217	0.339	0.026	0.051	-0.167	-0.100
Z6	-0.364	-0.027	0.299	0.035	0.192	0.115
Z7	-0.359	0.017	0.208	-0.024	0.295	-0.053
Z8	-0.317	-0.111	0.102	0.133	0.008	0.423
Z9	-0.191	-0.118	-0.164	0.321	-0.592	-0.073
Z10	-0.347	0.159	-0.023	0.146	-0.158	0.096
Z11	-0.247	0.209	-0.104	0.358	-0.243	0.318
Z12	-0.100	-0.065	-0.598	-0.131	0.115	0.017
Z13	-0.327	0.019	-0.207	-0.069	0.215	-0.247
Z14	-0.282	-0.114	-0.439	-0.155	0.213	0.011
Z15	-0.151	0.427	-0.019	-0.153	-0.040	-0.162
Z16	-0.023	0.136	-0.026	-0.715	-0.300	0.529
Z17	-0.174	0.422	0.013	-0.032	0.071	-0.188

Variable	PC7	PC8	PC9	PC10	PC11	PC12
Z1	-0.139	0.598	-0.161	-0.091	0.227	-0.025
Z2	0.061	-0.078	-0.086	0.101	0.170	0.049
Z3	0.316	0.426	-0.257	-0.009	-0.031	0.509
Z4	0.232	0.155	-0.125	-0.140	0.274	-0.743
Z5	-0.277	0.397	0.273	0.319	-0.395	-0.007
Z6	0.185	0.296	0.171	-0.026	-0.040	-0.070
Z7	0.290	0.269	0.310	-0.277	-0.040	-0.009

(continues)

TABLE A.2 *(Continued)*

Variable	PC1	PC2	PC3	PC4	PC5	PC6
Z8	-0.130	0.120	-0.184	0.673	0.259	-0.100
Z9	0.580	-0.186	0.178	0.163	-0.128	-0.044
Z10	-0.238	-0.069	-0.375	-0.300	0.185	0.110
Z11	-0.381	0.003	0.197	-0.367	-0.208	-0.086
Z12	-0.160	-0.062	0.368	0.085	0.080	0.052
Z13	0.032	-0.097	-0.493	0.094	-0.632	-0.187
Z14	-0.051	-0.054	0.196	0.056	0.099	-0.013
Z15	0.099	-0.197	0.033	0.027	0.256	0.315
Z16	0.183	0.071	-0.057	-0.073	-0.203	0.026
Z17	-0.018	0.005	0.135	0.230	0.031	-0.110

Variable	PC13	PC14	PC15	PC16	PC17
Z1	-0.185	-0.112	-0.016	-0.005	-0.010
Z2	0.052	0.236	-0.757	0.100	-0.091
Z3	-0.019	-0.133	-0.089	-0.024	-0.018
Z4	-0.046	0.141	0.115	-0.012	-0.002
Z5	0.231	0.356	0.172	-0.090	0.110
Z6	0.042	0.114	0.069	-0.197	-0.714
Z7	0.024	0.102	0.003	0.316	0.544
Z8	-0.164	0.060	0.006	0.130	0.193
Z9	0.115	0.026	0.047	-0.009	0.017
Z10	0.602	0.147	0.264	0.079	0.013
Z11	-0.414	-0.136	-0.181	-0.065	0.017
Z12	0.058	0.100	0.032	0.561	-0.293
Z13	-0.166	0.070	0.045	0.077	-0.021
Z14	0.171	0.086	-0.155	-0.692	0.220
Z15	-0.483	0.233	0.476	-0.123	0.003
Z16	0.051	-0.084	-0.015	0.021	0.022
Z17	0.189	-0.785	0.115	0.004	-0.008

TABLE A.3   Discriminant and Classification Analysis (MINITAB Output)

```
Linear Method for Response: C18
Predictors:
 X1 X2 X3 X4 X5 X6 X7 X8 X9 X10 X11 X12 X13 X14 X15 X16
X17

Group        1      2
Count      200     17

Summary of Classification

Put into         ....True Group....
Group                      1            2
1                        200            0
2                          0           17
Total N                  200           17
N Correct                200           17
Proportion             1.000        1.000

N = 217   N Correct = 217   Proportion Correct = 1.000

Squared Distance Between Groups
            1            2
1       0.0000      93.3935
2      93.3935       0.0000

Linear Discriminant Function for Group

                   1            2
Constant      -650.62      -679.79
X1               1.42         1.26
X2              -2.82        -3.12
X3              48.69        47.33
X4             145.05       145.52
X5               0.06        -0.06
X6               1.31         0.45
X7              -0.14         0.50
X8               0.34         0.37
X9               0.06         0.10
X10             -0.21        -0.01
X11              0.17         0.02
X12             -0.55        -0.77
X13             -0.08         0.02
X14              0.50         0.88
X15             26.10        46.00
X16              1.86         1.86
X17             -3.24        -3.58
```

TABLE A.4 Results of Stepwise Regression (MINITAB Output)

Alpha-to-Enter: 0.15 Alpha-to-Remove: 0.15
Reponse is C18 on 17 predictors, with N = 217

Step	1	2	3	4	5	6	7
Constant	1.9325	1.5151	1.3833	0.9777	1.0190	1.0992	1.0749
X5	-0.00157	-0.00119	-0.00147	-0.00133	-0.00127	-0.00131	-0.00124
T-Value	-15.15	-14.59	-18.68	-17.15	-17.03	-17.15	-15.63
P-Value	0.000	0.000	0.000	0.000	0.000	0.000	0.000
X13		0.00214	0.00162	0.00114	0.00091	0.00094	0.00087
T-Value		13.37	10.59	6.85	5.54	5.71	5.30
P-Value		0.000	0.000	0.000	0.000	0.000	0.000
X15			0.262	0.260	0.203	0.195	0.192
T-Value			8.30	8.84	6.66	6.37	6.36
P-Value			0.000	0.000	0.000	0.000	0.000
X14				0.00195	0.00184	0.00187	0.00321
T-Value				5.67	5.60	5.73	5.10
P-Value				0.000	0.000	0.000	0.000
X-10					0.00135	0.00157	0.00143
T-Value					4.78	5.20	4.72
P-Value					0.000	0.000	0.000
X8						-0.00063	-0.00085
T-Value						-1.96	-2.59
P-Value						0.051	0.010
X12							-0.00131
T-Value							-2.47
P-Value							0.014

	Step 8	Step 9	Step 10	Step 11	Step 12	Step 13	Step 14
S	0.188	0.139	0.121	0.113	0.108	0.107	0.106
R-Sq	51.63	73.65	80.09	82.71	84.40	84.68	85.12
R-Sq (adj)	51.41	73.41	79.81	82.39	84.03	84.25	84.62
C-p	538.4	198.3	100.2	61.5	37.3	34.9	30.2
Step	8	9	10	11	12	13	14
Constant	1.030	1.064	1.107	1.108	1.130	1.166	1.175
X5	-0.00122	-0.00120	-0.00120	-0.00118	-0.00120	-0.00118	-0.00119
T-Value	-15.26	-14.72	-14.80	-14.40	-14.87	-15.01	-15.34
P-Value	0.000	0.000	0.000	0.000	0.000	0.000	0.000
X13	0.00087	0.00086	0.00088	0.00086	0.00088	0.00088	0.00088
T-Value	5.28	5.23	5.37	5.24	5.40	5.54	5.58
P-Value	0.000	0.000	0.000	0.000	0.000	0.000	0.000
X15	0.195	0.200	0.190	0.178	0.175	0.155	0.155
T-Value	6.44	6.60	6.12	5.59	5.51	4.94	4.94
P-Value	0.000	0.000	0.000	0.000	0.000	0.000	0.000
X14	0.00325	0.00315	0.00325	0.00285	0.00204	0.00167	0.00171
T-Value	5.17	5.01	5.15	4.22	6.19	4.94	5.20
P-Value	0.000	0.000	0.000	0.000	0.000	0.000	0.00
X10	0.00136	0.00163	0.00170	0.00171	0.00182	0.00197	0.00198
T-Value	4.44	4.59	4.77	4.81	5.28	5.81	5.88
P-Value	0.000	0.000	0.000	0.000	0.000	0.000	0.000

(continues)

TABLE A.4 (*Continued*)

X8	-0.00090	-0.00081	-0.00081	-0.00086	-0.00075	0.00022
T-Value	-2.74	-2.42	-2.42	-2.58	-2.31	0.52
P-Value	0.007	0.016	0.016	0.011	0.022	0.600
X12	-0.00141	-0.00125	-0.00125	-0.00082		
T-Value	-2.64	-2.32	-2.31	-1.37		
P-Value	0.009	0.021	0.022	0.172		
X9	0.00024	0.00028	0.00034	0.00036	0.00035	0.00035
T-Value	1.47	1.71	2.04	2.12	2.05	2.16
P-Value	0.144	0.089	0.043	0.035	0.042	0.032
X11		-0.00117	-0.00137	-0.00156	-0.00181	-0.00185
T-Value		-1.49	-1.72	-1.95	-2.32	-2.46
P-Value		0.138	0.087	0.053	0.021	0.015
X1			-0.00126	-0.00133	-0.00137	-0.00134
T-Value			-1.57	-1.67	-1.71	-1.73
P-Value			0.117	0.097	0.089	0.085
X7				0.00095	0.00130	0.00483
T-Value				1.63	2.48	4.74
P-Value				0.105	0.014	0.000
X6						-0.0051
T-Value						-4.27
P-Value						0.000

S	0.105	0.105	0.105	0.104	0.104	0.102	0.101
R-Sq	85.27	85.43	85.60	85.78	85.65	86.50	86.48
R-Sq (adj)	84.71	84.79	84.90	85.02	84.96	85.77	85.82
C-p	29.8	29.4	28.7	27.8	27.8	16.8	15.0

Step	15
Constant	1.141
X5	−0.00114
T-Value	−14.62
P-Value	0.000
X13	0.00083
T-Value	5.33
P-Value	0.000
X15	0.153
T-Value	5.00
P-Value	0.000
X14	0.00348
T-Value	5.29
P-Value	0.000
X10	0.00177
T-Value	5.25
P-Value	0.000

(continues)

TABLE A.4 *(Continued)*

X12	−0.00191
T-Value	−3.09
P-Value	0.002
X9	0.00038
T-Value	2.38
P-Value	0.018
X11	−0.00127
T-Value	−1.67
P-Value	0.096
X1	−0.00125
T-Value	−1.65
P-Value	0.101
X7	0.0053
T-Value	5.22
P-Value	0.000
X6	−0.0069
T-Value	−5.28
P-Value	0.000
S	0.0994
R-Sq	87.08
R-Sq (adj)	86.39
X-p	7.7

TABLE A.5   Multiple Regression Analysis (MINITAB Output)

`Regression Analysis: Y versus X1, X2,...`

```
The regression equation is
Y = 1.17 - 0.00143 X1 - 0.00273 X2 - 0.0128 X3 + 0.0045 X4 - 0.00107 X5
    -0.00802 X6 + 0.00596 X7 + 0.000343 X8 + 0.000390 X9 + 0.00186 X10
    -0.00139 X11 - 0.00203 X12 + 0.000854 X13 + 0.00350 X14 + 0.186 X15
    -0.00006 X16 - 0.00318 X17
```

Predictor	Coef	SE Coef	T	P
Constant	1.1736	0.2469	4.75	0.000
X1	−0.0014339	0.0008149	−1.76	0.080
X2	−0.002734	0.004017	−0.68	0.497
X3	−0.01278	0.02656	−0.48	0.631
X4	0.01278	0.04587	0.10	0.922
X5	0.00447	0.00009411	−11.42	0.000
X6	−0.00107491	0.001746	−4.59	0.000
X7	−0.008015	0.001221	4.88	0.000
X8	0.005959	0.0004179	0.82	0.412
X9	0.0003433	0.0001646	2.37	0.019
X10	0.0003897	0.0003650	5.10	0.000
X11	0.0018598	0.0007864	−1.77	0.078
X12	−0.0013927	0.0006472	−3.13	0.002
X13	−0.0020250	0.0001616	5.28	0.000
X14	0.0008538	0.0006675	5.24	0.000
X15	0.0034999	0.04709	3.94	0.000
X16	0.18576	0.003012	−0.02	0.984
X17	−0.000060	0.008039	−0.40	0.693
	−0.003177			

`S = 0.1004  R-Sq = 87.2%  R-Sq(adj) = 86.1%`

Analysis of Variance

Source	DF	SS	MS	F	P
Regression	17	13.66102	0.80359	79.67	0.000
Residual Error	199	2.00718	0.01009		
Total	216	15.66820			

TABLE A.6   Neural Network Analysis (MATLAB Output)

```
» cd ann
» load input.txt-ascii
» load output.txt-ascii
» load test. txt-ascii
» input=input';
» output=output';
» test=test';
» [R,Q]=size(input);
» [S2,Q]=size(output);
» S1=10;
» net=newff (minmax(input),[S1 S2],{'tansig' 'tansig'},
'traingdx');
» net.trainParam.goal=0.1;
» net.trainParam.epochs=500;
» P=input;
» T=output;
» [net,tr]=train(net,P,T);
TRAINGDX, Epoch 0/500, MSE 0.472508/0.1, Gradient 109.592/
1e-006
TRAINGDX, Epoch 23/500, MSE 0.0873314/0.1, Gradient 73.7544
/1e-066
TRAINGDX, Performance goal met.

» P=test;
» A=sim(net,P);
» A

A =
  Columns 1 through 7
   0.1818 -0.0386 0.1472 0.0253 0.2754 0.3208 -0.3720
  Columns 8 through 14
   0.2614 0.0118 0.4912 0.2290 -0.1059 0.3299 0.2192
  Columns 15 through 21
   0.2213 0.1938 0.2769 0.1427 0.1629 0.2264 0.4081
  Columns 22 through 28
   0.2099 -0.0233 0.3312 0.2090 0.2813 0.0407 0.1086
  Columns 29 through 35
   0.1518 0.3605 -0.0494 0.0992 0.3274 0.3461 0.2328
  Columns 36 through 42
   0.0528 0.4984 0.2669 0.1319 0.4601 0.3381 -0.3734
  Columns 43 through 49
   0.1322 0.0310 0.2213 0.4409 0.4443 0.0755 -0.2122
  Columns 50 through 56
   -0.3009 0.5713 0.6330 0.7978 0.7943 0.9474 0.9171
  Column 57
   0.8707
```

TABLE A.7   Variables for Auto Marketing Case Study

1 Marital status of respondent
2 Sex of respondent
3 Age of respondent
4 Total # of people in household
5 Total number of children (computed)
6 Income group
7 How a veh looks is important—pers. view
8 Take srsly what people say—pers. view
9 Dont favor particular make—pers. view
10 Like to get make never had—pers. view
11 Just transportation—pers. view
12 Veh makes statement about you—pers. view
13 Only buy veh w/good mpg—pers. view
14 Would switch due to treatment—pers. view
15 Styling is important—pers. view
16 Buy easiest to maintain veh.—pers. view
17 Keep veh as long as possible—pers. view
18 Look for vehicle with power—pers. view
19 Like veh that are unique—pers. view
20 Serv warrants buy same make—pers. view
21 Put up w/impol salespeople—pers. view
22 Buy veh that are fun to drive—pers. view
23 Always buy same brand—pers. view
24 Consider myself an advocate—pers. view
25 Riding comfort—purch. reason
26 Gas mileage—purch. reason
27 Interior styling—purch. reason
28 Resale value—purch. reason
29 Int rates/credit/rebates—purch. reason
30 Passenger seating—purch. reason
31 Exterior styling—purch. reason
32 Price or deal offered—purch. reason
33 Warranty coverage—purch. reason
34 Technical innovations—purch. reason
35 Fun to drive—purch. reason
36 Well-made vehicle—purch. reason
37 Dealer service—purch. reason
38 Quietness—purch. reason
39 Safety features—purch. reason
40 Prestige—purch. reason
41 Value for the money—purch. reason
42 Power and pickup—purch. reason
43 Cost of serv and repairs—purch. reason
44 Cargo space—purch. reason
45 Good engine/trans—purch. reason

(continues)

TABLE A.7 (*Continued*)

46 Int roominess—purch. reason
47 Advice of relatives/frnd—purch. reason
48 Previous exper.w/model—purch. reason
49 Manufacturers reputation—purch. reason
50 Conven.dealer location—purch. reason
51 Durability (long lasting)—purch. reason
52 4-whl drive availability—purch. reason
53 Reliability (dependable)—purch. reason
54 Rear-whl drive available—purch. reason
55 Frnt-whl drive available—purch. reason

INDEX

Abnormalities, direction of, 43–50
  with both smaller-the-better and
    larger-the-better type
    variables, 46–48
  Gram–Schmidt process for
    identification of, 44–49
  with larger-the-better type
    variables, 45–46
  with smaller-the-better type
    variables, 48–49
  student admission system
    application of, 50–52
Abnormals, normals vs., 14–15
Additional information, test of, *see*
  Test of additional information
Adjoint matrices, 133, 140
Adjoint matrix method, 126–142
  and adjoint of a square matrix,
    128
  and cofactor computing, 127
  and determinant of a matrix,
    126–127
  examples of, 130–142
  for handling multicollinearity,
    128–129
Adjustment method for small
    correlations, 139, 141–144
American case studies, 177–191
  ASQ research fellowship grant,
    187–190
  auto marketing, 177–181
  gear motor assembly, 181–188

  transmission inspection system,
    190–191
American Supplier Institute (ASI),
  52
Artificial neural networks (ANN),
  13, 169–174
ASI (American Supplier Institute),
  52
ASQ research fellowship grant
  study, 187–190
Auto marketing case study, 177–
  181
  Mahalanobis space, construction
    of, 178–179
  measurement scale, validation
    of, 179
  useful variables, identification
    of, 179–181

Case studies:
  American, 177–191
  Japanese, 191–197
Categorical data, MTS/MTGS
  methods with, 85–95
  construction of Mahalanobis
    space, 89, 90
  data collection format for, 86
  description of variables, 88, 89
  identification of useful
    variables, 92–94
  sales and marketing application
    of, 87–95

Categorical data, MTS/MTGS
    methods with (*continued*)
    selection of variables, 88
    stages in, 86, 87
    validation of measurement
        scale, 89, 91–92
Classification:
    measurement vs., 14
    methods of, 11
Cofactor matrices, 126–128
Construction of Mahalanobis
    space:
    auto marketing case study, 178–
        179
    for categorical data, 89, 90
    gear motor assembly case study,
        184–185
Control charts, 169
    multivariate, 12–13, 169
Control factors, noise vs., 9
Correlation matrix, 131, 138
Covariance, 6
Customer quality, 8
Cusum charts, 13

Data analytic, probability-based
    inference vs., 15
Decision making, 3–4
Dimensionality reduction, 15, 61–
    81
    orthogonal arrays for, 62–63,
        92–94
    signal to noise (*S/N*) ratio for,
        63–81
Discriminant analysis, 11
Discriminant function, 11
Discrimination and classification
    method, 157–158
Distance measure, 6
Dynamic type *S/N* ratios, 65
    equations for (MTGS method),
        69–71
    equations for (MTS method),
        66–68
    as measure of functionality of
        system, 79

as simple measure to identify
    useful variables, 74, 77–78

ECM (expected cost of
    misclassification), 11
ED, *see* Euclidean distance
Engineered, on-line quality, 8
Engineering, quality, 10
Euclidean distance (ED), 21–22
Expected cost of misclassification
    (ECM), 11

Fasting plasma glucose (FPG),
    case study involving
    prediction of, 196–197
Feed-forward (backpropagation)
    ANN method, 169–174
Fisher's discriminant function,
    158–159
Ford Motor Company, 177
FPG, *see* Fasting plasma glucose
F-random variable, 12
F-ratio, 12, 15
Functionality, signal to noise
    (*S/N*) ratio as measure of
    system, 61, 79–81
Future research, recommendations
    for, 205

Gear motor assembly case study,
    181–188
    apparatus, 181–182
    characterization, 183–184
    high-resolution encoder, 182–
        183
    life test, 183
    Mahalanobis space, construction
        of, 184–185
    MTS scale, validation of, 185–
        186
    sensors, 182
    useful variables, selection of,
        186–187
Goal, 3
Graduate admission system
    (example), 50–52

Gram–Schmidt orthgonalization
  process (GSP), 24–29. *See
  also* MTGS method
  identifying direction of
    abnormals with, 44–49

Historical data, selection of, 147,
  149

Ichibiki & Co. Ltd., 191
Ideal function, 9
Ideal performance, 8
Input conditions, 6
Inverse correlation matrix method,
  see MTS method
Inverse matrices, 126, 128, 132

Japanese case studies, 191–197
  fasting plasma glucose (FPG),
    prediction of, 196–197
  soy sauce, utility rate of
    nitrogen while brewing,
    191–194
  water emulsion, measurement of
    oil in, 194–195

Larger-the-better type S/N ratios,
  65
  equations for (MTGS method),
    68–70
  equations for (MTS method),
    65–67
  as measure of functionality of
    system, 79
  as simple measure to identify
    useful variables, 71–78
Limitations, 205
Loss function approach, 15

Mahalanobis, P. C., 6
Mahalanobis distance (MD), 21–
  34, 36, 37, 39
  abnormalities distinguished by,
    43

calculation of, without
  assumption of distribution
  of variables, 27–29
for construction of measurement
  scale, 119–120
discriminant function, 11
Gram–Schmidt process for
  calculation of, 24–29
inverse matrix method for
  calculation of, 24
in multidimensional systems, 6–
  8
multiple, 142, 144–149
scaled, 8
and steps in MTGS, 31–33
and steps in MTS, 30–31
Mahalanobis space (MS), 7, 23
  in adjoint matrix method, 130
  auto marketing case study, 178–
    179
  calculation of mean of, 27–29
  construction of, for categorical
    data, 89, 90
  construction of measurement
    scale with, 30, 31
  gear motor assembly case study,
    184–185
  selection of, from historical
    data, 147, 149
Mahalanobis–Taguchi–Gram–
  Schmidt (MTGS) method,
  14–16, 21, 22
  artificial neural networks vs.,
    169–174
  with categorical data, 86–95
  direction of abnormalities
    calculation with, 43–50
  discrimination and classification
    method vs., 157–161
  feed-forward (backpropagation)
    ANN method vs., 169–174
  with medical diagnosis data,
    34–36
  and multicollinearity, 52–55
  multiple regression analysis vs.,
    165, 167–169

Mahalanobis–Taguchi–Gram–
    Schmidt (MTGS) method
    (*continued*)
  multivariate process control vs.,
      169
  objectives of, 23–29
  other methods of multivariate
      analysis vs., 153–174
  partial correlations with, 55–56
  principal component analysis
      vs., 155–157
  QLF for, 111–114
  role of OAs in, 62–63
  S/N ratio equations for, 68–71
  steps in, 31–33
  stepwise regression vs., 161–
      163
  subset selection in, 142
  test of additional information
      (Rao's test) vs., 163–169
  treatment of noise factors in,
      99–103
Mahalanobis–Taguchi System
    (MTS) method, 21, 22
  artificial neural networks vs.,
      169–174
  with categorical data, 86–95
  computing MDs in, 129
  and directions of abnormalities,
      43
  discrimination and classification
      method vs., 157–158
  feed-forward (backpropagation)
      ANN method vs., 169–174
  with medical diagnosis data,
      36–39
  and multicollinearity, 52–55
  multiple regression analysis vs.,
      165, 167–169
  multivariate process control vs.,
      169
  objectives of, 23–29
  other methods of multivariate
      analysis vs., 153–174
  principal component analysis
      vs., 155–157

  QLF for, 111–114
  role of OAs in, 62–63
  S/N ratio equations for, 65–68
  steps in, 30–31
  stepwise regression vs., 161–
      163
  subset selection in, 142
  test of additional information
      (Rao's test) vs., 163–167
  treatment of noise factors in,
      99–103
MATLAB, 171–174
MD, *see* Mahalanobis distance
Mean square deviation (MSD), 23
Measurement:
  classification vs., 14
  scale of, 6
  standard error and scale of,
      120–122
Medical diagnosis:
  backpropagation method with,
      171–174
  discriminant analysis of, 160–
      161
  multiple regression analysis
      with, 167, 168
  noise factors in, 100–103
  principal component analysis
      with, 155–156
  Rao's test, 164–169
  standard error for, 121–122
  stepwise regression analysis
      with, 162–163
  threshold in, 114–115
  use of MTGS and MTS
      methods with, 33–39
  variables in, 154
MLPs (multilayer perceptrons),
    170
MMD, *see* Multiple Mahalanobis
    distance
MR, *see* Multiple regression
MS, *see* Mahalanobis space
MSD (mean square deviation), 23
MTGS, *see* Mahalanobis–
    Taguchi–Gram–Schmidt
    method

MTS, *see* Mahalanobis–Taguchi System method
Multicollinearity, 52–55
    adjoint matrix method for handling, 128–130
Multidimensional system(s):
    correlations between variables in, 6
    definition of, 4
    Mahalanobis distance in, 6–8
    and robust engineering/Taguchi methods, 8–10
    typical, 5–6
    variables, 5–6
Multilayer perceptrons (MLPs), 170
Multiple Mahalanobis distance (MMD), 142, 144–149
Multiple regression (MR), 12, 165, 167–169
Multivariate diagnosis, 10–13
Multivariate process control, 169
Multivariate process control charts, 12–13
Multivariate systems, dimensionality reduction in, 61–81

Noise factors, 6, 99–103
    analysis combining different levels of, 102, 103
    control vs., 9
    MTS/MTGS treatment of, 99–103
    separate analysis of levels, 101
    as separate variable in analysis, 101, 102
    treatment of, 99
    unmeasurable, 103
Nonsingular matrices, 128
Normals, abnormals vs., 14–15

OAs, *see* Orthogonal arrays
Optimization, two-step, 9
Orthogonal arrays (OAs), 9
    for dimensionality reduction, 92–94

in MTS/MTGS, 62–63
in robust engineering, 62–63
Output, 6

Partial correlations, MTGS method with, 55–56
PCA, *see* Principal component analysis
PCs, *see* Principal components
Performance, ideal, 8
Principal component analysis (PCA), 10–11, 15, 155–157
Principal components (PCs), 155–157
Probability-based inference, data analytic vs., 15

QE (quality engineering), 10
Quadratic loss function (QLF), 15
    for determining threshold, 109–114
    for larger-the-better characteristic, 110
    in MTS/MTGS method, 111–114
    for nominal-the-best characteristic, 109–110
    for smaller-the-better characteristic, 111
Qualitative data, *see* Categorical data, MTS/MTGS methods with
Quality:
    customer, 8
    engineered, 8
Quality engineering (QE), 10

Rao's test, *see* Test of additional information
RE, *see* Robust engineering
Regression. stepwise, 11–12, 15
Robust engineering (RE), 8
    QLF in, 109
    role of OAs in, 62

Scaled Mahalanobis distance, 8
Scientific contributions, 203–204
Shewhart charts, 13
  multivariate, 169
Signal-to-noise (S/N) ratio, 9, 15, 23–24, 34, 63–81
  dimensionality reduction for, 63–81
  dynamic type of, 65–71, 74, 77–79
  equations for (MTGS method), 68–71
  equations for (MTS method), 65–68
  functionality of the system as measure of, 61
  identification of useful variables with, 92–95
  larger-the-better type of, 63–69, 71–76, 79
  as measure of functionality of system, 79–81
  to predict given conditions, 80, 81
  relationship between variability reduction and gain in, 79, 80
  as simple measure to identify useful variables, 71–78
Singular matrices, 126, 128
Small correlations, adjustment method for, 139, 141–144
S/N ratio, *see* Signal-to-noise ratio
Solution strategy, 16, 17
Soy sauce, case study of utility rate of nitrogen while brewing, 191–194
Square matrices, 126–128
Standard error of the measurement scale, 120–122
Stepwise regression, 11–12, 15
  MTS/MTGS methods vs., 161–163
Strategy, solution, 16, 17

Student admission system, threshold in, 115–116
Subset selection, using MMD method, 142, 144–149
System functionality, signal to noise (S/N) ratio as measure of, 61, 79–81

Taguchi Methods (TM), 8–10
Test of additional information (Rao's test), 12, 15, 163–169
Threshold, 107–116
  determination of, 112–114
  general, 108
  in medical diagnosis case, 114–115
  specific, 108
  in student admission system, 114–116
  when only good abnormals are present, 114
TM, *see* Taguchi Methods
Transmission inspection system case study, 190–191
Two-step optimization, 9

Unit distance, 30. *See also* Mahalanobis distance
Unit group, 7
Unit space, 23, 30, 31, 130. *See also* Mahalanobis space

Variability reduction, relationship between gain in S/N ratio and, 79, 80
Variables, correlations between, 6
Vivekananda, Swami, 3

Water emulsion, case study involving measurement of oil in, 194–195
Wayne State University, 177

Xerox Corporation, 181